About Island Press

Since 1984, the nonprofit Island Press has been stimulating, shaping, and communicating the ideas that are essential for solving environmental problems worldwide. With more than 800 titles in print and some 40 new releases each year, we are the nation's leading publisher on environmental issues. We identify innovative thinkers and emerging trends in the environmental field. We work with world-renowned experts and authors to develop cross-disciplinary solutions to environmental challenges.

Island Press designs and implements coordinated book publication campaigns in order to communicate our critical messages in print, in person, and online using the latest technologies, programs, and the media. Our goal: to reach targeted audiences—scientists, policymakers, environmental advocates, the media, and concerned citizens—who can and will take action to protect the plants and animals that enrich our world, the ecosystems we need to survive, the water we drink, and the air we breathe.

Island Press gratefully acknowledges the support of its work by the Agua Fund, Inc., Annenberg Foundation, The Christensen Fund, The Nathan Cummings Foundation, The Geraldine R. Dodge Foundation, Doris Duke Charitable Foundation, The Educational Foundation of America, Betsy and Jesse Fink Foundation, The William and Flora Hewlett Foundation, The Kendeda Fund, The Andrew W. Mellon Foundation, The Curtis and Edith Munson Foundation, Oak Foundation, The Overbrook Foundation, the David and Lucile Packard Foundation, The Summit Fund of Washington, Trust for Architectural Easements, Wallace Global Fund, The Winslow Foundation, and other generous donors.

The opinions expressed in this book are those of the author(s) and do not necessarily reflect the views of our donors.

Public Produce

Public Produce

The New Urban Agriculture

Darrin Nordahl

ISLANDPRESS

Washington | Covelo | London

ISLAND PRESS is a trademark of The Center for Resource Economics.

Library of Congress Cataloging-in-Publication Data

Nordahl, Darrin.
 Public produce : the new urban agriculture / Darrin Nordahl.
   p. cm.
 Includes bibliographical references and index.
 ISBN-13: 978-1-59726-587-4 (hardcover : alk. paper)
 ISBN-10: 1-59726-587-X (hardcover : alk. paper)
 ISBN-13: 978-1-59726-588-1 (pbk. : alk. paper)
 ISBN-10: 1-59726-588-8 (pbk. : alk. paper) 1. Urban agriculture—United States. 2. Food supply—United States. I. Title.
 S441.N77 2009
 338.1'91732—dc22

                                                        2009010262

Printed on recycled, acid-free paper

Manufactured in the United States of America
10 9 8 7 6 5 4 3 2 1

Keywords: agritourism; community health; Davenport, Iowa; Davis, California; food equity; food policy; food safety; food security; foraging; gleaning; public policy; public space; urban agriculture; urban design; urban farming

For Lara

Contents

Preface

2008 was a year many hope to forget, but it will likely remain burned in our memories. For the first time in our nation's history, gasoline prices exceeded four dollars a gallon across the country. And it was a time when the many loose threads of our economy seemed to simultaneously be pulled in every direction, unraveling the very fabric of our lives. In 2008, this nation witnessed millions of lost jobs and sobering levels of unemployment; the subprime mortgage crisis, the collapse of Fannie Mae and Freddy Mac, and the subsequent swelling rates of home foreclosures; the free fall of our most revered investment banks and retailers; the loss of billions of dollars in retirement accounts and investments, affecting countless people, from the middle-class working stiff to the *über*wealthy caught up in the Bernard Madoff Ponzi-scheme scandal; and the buckling of one the largest and greatest industries in the history of our nation, the Big-Three automobile manufacturers.

And then there was Mother Nature's wrath. In 2008, floods across the Midwest, drought along the coasts, ice storms, power outages, and weather anomalies across the Southeast and Northeast offered further proof that climate change is, indeed, real. Families were displaced, lives were lost, crops were ruined, and America's downfall in the global economy had devastating effects on developed and developing countries around the world.

Amid the economic and climatic turmoil, the price of our food and the numbers of this nation's hungry skyrocketed. The year 2008 kicked off the worst financial crisis since the Great Depression. And 2008 marked the first year since the 1930s that many Americans were truly beginning to wonder where their next meal would come from.

This book began years ago, in more prosperous and secure economic times, as a topic for my students at the University of California Berkeley Extension. Public produce then was simply an idea to showcase how public space and public policy could work together to reduce food insecurity for the destitute and the perennially hungry. But during the economic downturn that began to unfold in 2008, it quickly became apparent that people across the country, even the middle class, could soon be joining the ranks of our nation's most deprived.

Early in my research, many had argued that this idea of public produce to aid those truly afflicted by the rising cost of food is likely infeasible. And even if it were feasible, they insisted, it would only be remotely effective on the "Left Coast," where people are liberal and the climate mild. Since 2008, people's minds, like the climate, have been changing. In light of the seemingly daily headlines announcing the rising cost of produce, the weather aberrations and subsequent crop loss, the pathogen-infected produce, the falling out of favor of "industrial organic," and the insatiable demand for locally grown produce, folks are beginning to admit that a public network of food-growing opportunities could benefit more people than just the utterly impoverished. Though this book focuses heavily on social equity, aimed principally at those with little choice with regard to food, it is meant to illustrate that regardless of one's financial station in life, there are benefits, both individual and communal, to returning our urban lifestyle to its agrarian roots, and reinstating a modicum of self-sustenance.

Economies are usually cyclical. While circumstances today are dire, prosperity is out there somewhere on the horizon. But in the face of unprecedented global warming, those times of prosperity may be more sporadic and unpredictable than they were following the Great Depression. Climate change, as the

National Oceanic and Atmospheric Administration reported at the dawn of 2009, is irreversible. Well, at least for the next thousand years anyway.

With global warming comes an unstable climate, and with an unstable climate comes an unstable food supply. Petroleum prices will continue to increase, and our nation's food—which is inextricably linked to oil—will see price increases as well. Though oil prices fell shortly after they spiked in 2008, the relief is likely ephemeral. Until communities figure out how to provide for themselves, instead of relying on a handful of petrophilic agribusinesses in remote locations in our country and abroad, our satiety will be tenuous.

There is a good deal of focus on California and Iowa in this book, which can be expected, as they are the two states I call home. As such, they are where I have witnessed firsthand the many innovative ideas toward public food production. But I have also chosen to highlight these two seemingly different landscapes (and cultures) to help prove that food is the great equalizer. In the culinary world of *haute cuisine*, California and Iowa could not be more different. Yet, in light of the demand for fresh, wholesome food at an affordable price, expanding waistlines, crop failures from an eccentric climate, and increasing instances of poverty and hunger, Iowa and California may as well be conjoined. Though I sometimes think Iowa and much of the Midwest are ten years behind the progress and innovation being made on the coasts, the Midwest is, for once, actually keeping pace with—and in some cases exceeding—the pioneering policies that are being adopted along the typically more progressive edges of our country.

What has typically been a grassroots approach to food security (e.g., community activists lobbying local government officials to allow modest community gardens on vacant lands owned by the city) is now becoming an endeavor *initiated* by government staff. While this top-down approach to community food security is good news for food advocates, it is not particularly newsworthy. The topic of public produce—which can more descriptively be defined as municipal agriculture—does not receive a lot of publicity or fanfare, so it is difficult to unearth research on this topic. As such, many municipalities are implementing programs more or less from scratch. It is my intent to showcase a few innovative policies and implementation strategies that are currently happening across the country, to illustrate the breadth of innovation, provide a modest list of resources, and more importantly, further encourage thought and discourse on the subject. Though the municipal agriculture movement is nascent, it is burgeoning, moving quickly from ideation to palpability.

Much of the information I have gleaned comes from Internet research, word of mouth, and, most prevalent, direct observations of what communities are doing in the arena of municipal agriculture. There is not yet an abundant supply of published material dealing specifically with this topic. Yet, it seems that daily a new headline appears on the rising cost of food, pathogen outbreaks, obesity and diabetes, and the growing demand for local food options. I will venture a guess that the current paucity of published work devoted to the concept of public produce will soon be a thing of the past.

A little more than twenty years ago, a book could have been published that espoused the environmental benefits of recycling and urged municipalities to organize citywide recycling programs so that everyone in the community had the ability to lessen their ecological footprint. Such a book would be pointless today. Cities across the nation now realize the environmental good that comes with recycling. Public officials have figured out how to collect, sort, and recycle a variety of materials and how to effectively educate their citizens on what to recycle and why recycling is good for them and their community.

It is my sincerest hope that in twenty years, a book espousing municipal-organized agriculture will also be pointless. By that time, public officials across the nation will have implemented a variety of strategies to produce food throughout the city so that everyone in the community has the ability to eat healthy, whenever and wherever. They will have figured out how to grow, maintain, harvest, and process an abundance of fresh fruits and vegetables, while creating beautiful and inspiring edible landscapes. Programs will have been created to educate citizens about food and food choices, and why municipal agriculture, like recycling, is good for them and their community. In twenty years, *Public Produce* will be out of print, and there will be no reason for its resurrection.

Serendipity

A reward awaits those crossing the Potomac on the footbridge from Theodore Roosevelt Island to the George Washington Parkway. At least, that is how my friend and I thought of the lone apple tree on the western bank of the river. The two of us were headed back to the car after exploring the memorial island, feeling hot, tired, and a bit hungry. Apparently, others before us had discovered this treat, as most of the apples left on the tree were out of reach. We managed to grab one apple each, and, though they were still a couple of weeks from being ripe, the tart green fruit provided momentary satiety and invigorated our spirits.

That apple tree, a solitary symbol of an agrarian landscape in an otherwise intensely urban setting, caught us by surprise. Though we were getting hungry, we were not seeking food, especially within our immediate surrounds. After all, it is rare that one stumbles across fresh, free produce in the middle of a big city. For a city whose lore and landscape are so entwined with cherry trees, why something as seemingly innocuous as a fruit tree—any fruit tree—should provoke wonderment was a bit puzzling.

Perhaps we were caught unawares because, even in our nation's capital, where more than three-thousand iconic cherry trees have become one of the city's premier tourist attractions, we are accustomed to plants in the urban environment providing simple aesthetics, rather than wholesome nourishment. The Kwanzan cherry, the specific variety that makes up the bulk of the cherry trees in East Potomac Park, is a fruitless cultivar. The Yoshino cherry—the principal cultivar that encircles the Tidal Basin and punctuates the Washington Monument grounds—does produce fruit, though it is stony and unpalatable to all but birds. There is no denying the poetic beauty of these trees—a generous gift from Japan—whose showy blossoms are an allegory of friendship. Yet, I wonder, if flowers can be an accepted symbol of goodwill and inspire all who gaze upon them, can fruit become an accepted symbol of equity, for all to eat?

Three thousand miles west of that apple tree near Teddy Roosevelt Island, at the other end of U.S. 50, apple orchards command tourist attention. During the ripening months of September and October, throngs of urbanites retreat to an area known as "Apple Hill" simply for the opportunity to harvest fresh apples. These tourists travel to this Sierra Nevada locale from all over northern California, many from as far away as the Bay Area. That people are willing to drive 140 miles from San Francisco for the unique experience of picking apples off the tree is testament to how hungry urbanites are for a bit of agrarianism.

Apple Hill and other "U-pick" farms throughout the country are part of a fast-growing industry known as "agritourism." For an hour, a day, or a week, agritourism sites and excursions allow the urbanite to escape the trappings of city life, promising personal rejuvenation through the agrarian experience. Opportunities to pick fruits and vegetables, help work the land, taste fresh honey, milk, and eggs, or even crush grapes and make wine compel the agritourist.

Spending good money and free time on an agrarian experience might seem absurd to our forefathers. But the success of agritourism—its *raison d'être*, in fact—stems from a growing citizenry that has lived life never having plucked a berry from the bush or an apple from the tree. In a nation with such deep agrarian roots, it is almost inconceivable that today there would be such a chasm between the American family and the farm. But shortly after World War II, during the urban renewal of our inner cities and the sprawling development of our suburban settlements, the small family farms, public gardens, and individual produce markets and stands disappeared. And with the disappearance of these once-ubiquitous displays of food and food production, we forgot what was once common knowledge: where food comes from, how to grow it, and when it is ready to eat.

Like other land patterns in post–World War II city development, there was no longer room for farms or fussy edible landscapes. Cities were to be streamlined and compartmentalized, with the home, workplace, marketplace, and open space all separated from each other. The zoning that mandated the separation of land uses also prohibited agriculture within the more urbanized neighborhoods of the city. Once the land uses were separated and the impurities of agriculture removed, a new settlement was born, one that commanded cleaner landscaping: well manicured, sterile varieties of trees, shrubs, and ground covers.

Suburban sprawl picked up where zoning laws left off and pushed agriculture activities even further from the city center. Those farms not consumed by residential subdivisions became aggregated with other farms. As such, the second half of the twentieth century saw the number of farms in America dwindle from more than six million in 1940 to just two million at the dawn of the new millennium.[1]

And so, the agricultural paradigm had shifted. The pervasive ideology of the mid-twentieth century became that food production was no longer suitable in and around our cities, as it had been for centuries. Growing fruits and vegetables was no longer the work of community-minded individuals and families on small local farms, but endeavors better suited to corporate-owned, factory-like "agribusiness" in more distant parts of the country.

Now, as the twenty-first century is underway, a cresting wave is readying the backlash against large-scale corporate agriculture on fields hundreds—if not thousands—of miles from where we live; against mass-produced, chemically grown produce; against the rising costs of food and the declining health of the American people. The organic movement is ceding to the "buy-local" movement; fast food has become a pejorative term,[2] while "slow food" seems to be the choice of the future. Farmers' markets, community-supported agriculture groups (CSAs), and small produce stands are part of a burgeoning system of local agriculture that is enjoying a popularity not witnessed in more than half a century. And the time is ripe to explore how we can expand this network of local food options to meet the growing demand of consumers by bringing agriculture back into our cities.

This book explores the role of food-growing opportunities in the development of our cities, and the options of gathering food from the urban environment. Admittedly, it is unrealistic to believe that in the near future Americans will only eat locally grown, seasonally available produce. We will still want bananas, oranges, and avocados even if we live in Wisconsin, or tomatoes, peppers, and corn in February, regardless of where we live. It is also unrealistic to assume that urban Americans will move to the countryside and completely return to subsistence agriculture. This book is about providing food *choices* within the city—where the majority of the U.S. population lives today, and with continued urbanization projected—and about how to achieve healthful, low-cost supplements to our diet. *Public Produce* examines local food options through the lens of social equity: closing the food gap between the inner-city poor (and increasingly the lower-middle and middle class) and the high prices of supermarket organic and farmers' market produce; improving the health of the American population, especially our children, who increasingly lack everyday accessibility to fresh produce; setting aside land so that apartment and condominium dwellers—who have little, if any, land to sow—can have the same opportunity to grow gardens as homeowners; providing a sense of self-sufficiency to even the well-to-do by giving them an opportunity to forage; and recognizing the social relationships and prosperous citizenry that could result if city spaces could help provide food for all.

Toward the goal of food justice, this book is specifically about fresh produce grown on public land, and thus available to all members of the public—for gathering or gleaning, for purchase or trade. And, because this food is grown on public land, this book examines the efforts, programs, and policies that are being ushered and implemented by local governments. If a network of locally available, publicly accessible produce is to be successful, the largest single landowner within the city—the municipality itself—will have to be engaged.

The idea that municipal government should shape the food supply in the face of increased costs is not a new idea. In 1977, the City of Hartford, Connecticut, devised an initiative to address the increasing cost of food available to the city's lower-income residents. Hartford's initiative was in direct response to the fecklessness of the federal programs available at the time—namely the Food Stamp Program and School Breakfast Program. The initiative focused on food-distribution projects that not only helped close the food gap between supermarket produce and the inner-city poor, but provided an economic boost to Connecticut's farmers, butchers, canners, and other food processors. But Hartford's initiative also focused on the establishment of facilities to nurture a system of public produce within the city itself: community and youth gardens, solar greenhouses, cold frames, and rooftop production—all on land and within structures provided by the municipality.[3]

At the heart of these pleas for a more equitable system of food production is *food security*: daily access to an adequate supply of nutritious, affordable, and safe food. The recent outbreaks of *Escherichia coli* (*E. coli*) infecting spinach from California and *Salmonella* contaminating peppers packed in Texas, peanuts in Georgia, and pistachios in California, reveal that our fresh-produce farms and distribution centers may not be as safe and sterile as we thought. Climate change that is producing drought in California and Florida, at the same time as flooding in Iowa, is reducing crop yields. Pest infestations are reducing crop yields as well. California is grappling with the light brown apple moth, a problematic little bugger that has fluttered into the state from Australia. Its diet not only consists of apples, but of almost every crop the state produces. Florida Department of Agriculture spokesman Terence McElroy notes, "Our office is getting reports of at least one new pest or disease of significant economic concern per month."[4] California and Florida provide much of the variety of our nation's fresh fruits, vegetables, and nuts. It is unfortunate that relatively isolated agricultural problems in a couple of states are felt nationwide, but such is the nature of our current food-supply system. As a measure of insurance, this is perhaps reason enough to employ a more local, public system of food production.

Weather anomalies, pest infestations, and bacterial contaminations obviously limit the food supply, which in turn drives up prices. But there is another, perhaps more pervasive, reason for the recent spike in the cost of fresh produce: oil. The large-scale, specialized agribusinesses that furnish much of the food in the United States rely heavily on oil. Idaho produces much of the nation's potatoes; Washington our apples; Michigan our blueberries; California our broccoli; and Iowa our corn and soybeans (the bulk of which is consumed by livestock, or processed into corn syrup, ethanol, and partially-hydrogenated oils or transfats). The soaring cost of oil affects these large-scale agricultural endeavors in many interconnected ways: from the fuel used to power the tractors and combine harvesters, to the petroleum-based herbicides and pesticides liberally sprayed on the fields, and back to the fuel used to power the diesel trucks that deliver the produce hundreds, if not thousands, of miles to our urban markets.

That increasing distance to market—measured in "food miles"—is of great concern in the face of a shrinking oil supply and its ever-rising cost. If our produce only came from within our nation's boundaries, perhaps those food miles could be manageable. We now import considerable produce from large, multinational food conglomerates in countries like Canada, Mexico, Chile, and, increasingly, New Zealand and China. As it is, the average produce item in our supermarkets comes from more than 1,500 miles away.[5] As food producers become bigger and more specialized, their distances away from cities become greater and energy consumption increases. Reduce the distance an apple travels from the tree to your hand, and a reduction in price could result.

The people most affected by the rising cost of produce are low-income individuals, as well as single-parent and single-income families. Foodborne pathogens and pests, on the other hand, affect everyone, regardless of financial position, and have become a national security concern. Hunger is obviously the result of food insecurity, and can be seen the world over. But in America, *obesity* is a food security issue as well. The farmer is no match for the deep-pocket marketing campaigns of our fast-food chains and processed-food conglomerates, especially during tough economic times. (Amid our current fiscal free fall, McDonald's continues to post strong earnings.)[6] Children are especially susceptible to advertising, a fact marketing consultants use to their advantage. To combat the salvo of fast food and processed-food commercials, signs, and billboard advertisements, fresh, whole foods ought to be equally omnipresent in our urban environment, to remind children—everyone, for that matter—of healthful food alternatives. We thus have to change the way we think about plants and public spaces in the urban environment—not as merely providing aesthetic and recreational value, but sustenance and nutrition as well.

Addressing food security is reason alone to explore the notion of a more public system of food production, but there are certainly more. As will become evident in subsequent chapters, public produce is helping to attain broad civic aims, such as providing small-business financial assistance; boosting civic pride and building community; reducing crime; strengthening our connection to place; and reintroducing seasons and the natural cycles of life to our young and not-so-young. In short, food choices found in our urban surrounds can give citizens a more bountiful life.

There are also environmental benefits. Mayor Richard M. Daley is leading the way in creating a more environmentally friendly city through the greening of public spaces in Chicago. The Windy City offers an inspirational success story, rocketing from what many people thought was a fallen, dilapidated city to one of the greenest and greatest cities in the nation. Daley's investment in the environment has not only improved the ecology and aesthetics of Chicago, but has brought in billions of tourist dollars, triggered a spike in development interest, and garnered the attention of civic leaders and city builders around the globe.

The physical greening of Chicago—through the planting of countless trees, shrubs, and perennials along the streets, atop roofs, within parks, plazas, and other public spaces of Chicago—is certainly praiseworthy. However, the next evolution of greening our civic spaces should focus on the value each tree, shrub, and perennial provides to the public. An elm tree, for example, offers beauty and shade, providing a natural, fossil-free source of air conditioning. It also sequesters carbon dioxide emissions from the environment and replaces them with oxygen; creates habitat for birds and other urban wildlife; and reduces storm-water runoff. An elm tree also helps give scale and interest to the street, enhances buildings, and as such, raises property values. But can a tree do all of these things and go one step further, by providing food for human consumption as well? Adding food to the list of benefits a public tree can provide

greatly increases its value to the city's citizens and visitors. Landscape architects, as designers of our urban public spaces, have proven adept at using plants
to address concerns of comfort, maintenance, aesthetics, and other socio-
environmental factors. Adding food to that list is well within their regimen, and
something that should be demanded by clients.

The best place to realize the environmental, economic, and equitable benefits of a more local system of agriculture may not be in some rural or exurban
location, but in and among the places we pass by daily on our way to work,
home, school, commerce, and recreation. It may not seem so to the casual observer, but when the sum of all the public spaces in a typical city is figured, the
municipality itself is the largest single landlord. The sheer abundance of land
within public control necessitates a hard look at how it can best serve the needs
of its shareholders. This could mean the land needs to be as productive as bio-
logically possible, that every square foot has value to those who use it or pass by.
Plazas, parks, town squares, city streets, and the grounds around our parking
lots, libraries, schools, city halls, and courthouses are prime locations to consider when rethinking the role of public space in our cities—and how to add
value to those spaces if they are currently underutilized. Hence, the efforts pro-
filed in this book go beyond the mere greening of our city spaces; they illustrate
how public space can produce a commodity that can be consumed by the human end user, namely, food.

More than just a discussion of providing places for the occasional community garden, the intent of this book is to examine how the intricate web of public
space within cities can be used for more prolific food production. Many of the
strands that frame this web will be in the form of community gardens, but not
those relegated to the vacant lot in a distressed neighborhood, or tucked away in
the back corner of a little used park. Rather, cities in the near future may be
looking at community gardening on the magnitude of that now being implemented in Detroit. Often cited as the exemplar of urban community gardening,
the City of Detroit has demonstrated what municipalities can do to return
blighted and abandoned land to productive use. Of course, the motivation for
such greening efforts is driven by the economic turmoil of thousands of vacant
lots under city ownership. Nevertheless, the city's efforts are inspiring. Public

officials are making the best out of a difficult financial situation, and achieving communal and environmental benefits in the process.

Beyond community gardening at the scale of that in Detroit, this is a critical examination of *all* the plants in *all* the public spaces within the city: fruit trees and shrubs along streets and in medians; orchards in parks; herbs and vegetables in planters located on plazas and sidewalks in our commercial areas; and roof-top agriculture, to name a few. Most notably, this book scrutinizes the dense, multi-stranded network of food-growing opportunities accessible to the public that could be realized with the active support and involvement of city government.

Some government officials already recognize the dire need for municipalities to engage in food-producing alternatives. Susan Anderson, senior horticulturalist for the City of Davenport, Iowa, argues that one responsibility of local government "falls in the area of dedication of land and management of it for the common good. In an urban environment how do we provide the opportunity for people to access land they can use for food?" Anderson uses her home city of Davenport as an example: "We are an urban community. Preserving agricultural land as a resource is important but in an urban setting commercial, large-scale farming operations of the Midwest variety aren't going to help someone downtown."[7]

Anderson believes local government should set aside public land expressly for the purpose of urban gardening. She further contends that such land dedication and management "becomes a wellness issue for the community. Actually, it is very attractive to those of us who are concerned with the quality of our soils, depletion of minerals and nutrients essential to healthy people and plants, to see a community that provides access to locally grown, fresh food sources and/or the ability to create our own."[8]

The system of municipal agriculture that Anderson describes could be a manifestation of what the late Thomas Lyson, a distinguished professor in the Department of Development Sociology at Cornell University, called "civic agriculture." According to Lyson, civic agriculture "embodies a commitment to developing and strengthening an economically, environmentally, and socially sustainable system of agriculture and food production that relies on local resources and serves local markets and consumers. The imperative to earn a profit is filtered through a set of cooperative and mutually supporting social relations. Community problem solving rather than individual competition is the foundation of civic agriculture."[9]

By contrast, Lyson argued, "Large-scale, absentee-owned, factory-like fruit and vegetable farms that rely on large numbers of migrant workers and sell their produce for export around the world would not be deemed very civic."[10]

Lyson recognized a growing hunger for civic agriculture, as evidenced by the popularity of farmers' markets, community-supported agriculture groups, and community gardens throughout the country. Farmers' markets increased from 1,750 in 1994 to nearly 4,400 today. CSAs were virtually nonexistent in the 1980s; now there are more than 1,200.[11] The *slow food* movement (as opposed to fast food) is garnering interest throughout the country. The term *locavore* now appears in *The New Oxford American Dictionary*; it means someone who only eats what is grown or produced locally, usually within a hundred-mile radius.[12] Eating only that food which is produced within a hundred miles of your dinner plate is an admirable challenge given today's methods of food production, but it is a distance that would seem formidably long to our grandparents and great-grandparents. For them, the thought of carrots, tomatoes, onions, or potatoes traveling 100 miles to consumers would be unfathomable.

In addition to advocating for smaller, independent farms located closer to cities, planners, environmentalists, policy makers, and educators are also urging the preservation of existing agricultural land within the city boundary, and, in some cases, new farms interwoven into the urban fabric. Unfortunately, it may be too late for this latter policy reform in some cities. Some metropolises are just too big to have farms very near the principal city's center. For those older, denser cities along the Mid-Atlantic, for example, there is little to no agricultural land left within or around the city to preserve; and little room for new farms. Nor would this policy benefit inner-city or downtown residents in most urban communities, as Susan Anderson noted. For these communities, the only option for local food production may be to explore the available, arable land within their urban environs as an opportunity to establish a vast network of small-scale, yet abundant food-producing activities.

Of course, many hurdles lie in the way of providing a healthier, more equitable urban landscape. One of the tallest may be our newly gained ignorance of food. For any system of public produce to be successful, we as a nation will have to reeducate ourselves about food, what it looks like, where it grows, and when it is ready to harvest. In short, we need to get back to our agrarian roots. I have witnessed adults convinced that pineapples grow on trees. A very young, very naïve vegan acquaintance once explained she could not have coconut milk because she gave up meat and dairy in her diet. I have been a member of a well-intentioned CSA that did not know the proper time to harvest okra. What were delivered were large pods the size of Anaheim chili peppers. (When okra pods are allowed to grow large their flesh becomes tough to the point of inedibility.)

Even people living in rural areas are no more food-savvy than the typical big-city dweller; ironically, less so. I worked with a young woman from rural Iowa who had never tasted eggplant, and admitted she probably would not be able to recognize one. (Eggplant, incidentally, grows quite well in Iowa.) My neighbor recounted a conversation she had with a teenager who loved to eat guacamole, but had never seen an avocado. Many Iowans never have eaten tofu, tasted soy milk, or ever heard of edamame—yet Iowa is the largest grower of soybeans in the country. On a recent educational tour of Davenport's conservatory that was aimed at teaching children about food and plants, one child asked if the sunflowers would grow "ranch-flavored" seeds. Another asked if the small oranges on a tree in the grow house were pumpkins or watermelons. As Susan Anderson lamented, "Do you think kids in Iowa know where their food comes from? These kids live in Davenport, not Chicago. They can drive five miles and see a farm. On a good day they can smell the corn-processing plants. There is a huge disconnect going on. Unless we work to initiate a process for change it won't get better."[13]

If public produce is to succeed in our cities, educational programs are needed to reacquaint us with food, to help us recognize which plants are edible and which are ornamental, and to teach us how to plant, how to care for, and how to harvest food. In short, we have much to relearn about food and agriculture as we explore opportunities for them in our urban settings.

Thankfully, we can learn from the various bits of municipally organized urban agriculture on public land that are already happening across the country. This book highlights a few of those efforts. Many of these efforts are, admittedly, small in scope. Collectively, they indicate a budding shift in public policy taking root throughout the country. Urban agriculture on public land, though currently in an embryonic state, is certainly real. The collection of assorted, independent examples underway throughout American cities big and small offers a glimpse of a trend driven not by a central government policy, but by a local one, and the communities' desire for more economically viable, environmentally sustainable, locally available, and healthful choices in food production. These efforts are varied: restaurateurs seeking to reduce overhead costs by growing their own produce, or being willing to trade for it; city officials, both hired and elected, using public space under their management for the production of food; school grounds being replanted with edible gardens to help teach children about where food comes from and how to grow it (and to entice them to eat healthier); neighborhood groups organizing community gardens and promoting usufruct laws (the legal right to harvest fruit belonging to a private party if it overhangs, or is accessible from, public property); and the rise of "guerilla gardeners"—vigilante groups that take over vacant or blighted land in the city and return it to productivity and beauty through the planting and management of gardens. Though their work is benign and their mission inspirational, there is reason for their "guerrilla" moniker: their tactics border on extremism. Regardless, their actions and the others mentioned point out the lengths to which citizens are going to increase accessibility of fresh produce.

It is time for municipal government to recognize these urban food-producing endeavors, embrace them, help manage, and even build upon them. Indeed, many of the grassroots efforts are initiated by government employees themselves—dedicated civil servants bent on improving the quality of the city and the quality of life for its inhabitants. Though each of these examples of public produce is indeed small and independent, their collective sum illustrates that there is both a need and a desire to supplement our existing food-production methods outside the city with opportunities within the city itself. Working in concert, each venture—regardless of size or scope—contributes to making fresh produce more available to the public. And, in so doing, each can help reinforce a sense of place and build community; nourish the needy; provide economic assistance to entrepreneurs; promote food literacy and good health to all; allow for serendipitous sustenance; and add a bit of agrarianism back into our urbanism.

CHAPTER ONE

Food Security

To put it simply, Americans have been eating oil and natural gas for the past century, at an ever-accelerating pace. Without the massive "inputs" of cheap gasoline and diesel fuel for machines, irrigation, and trucking, or petroleum-based herbicides and pesticides, or fertilizers made out of natural gas, Americans will be compelled to radically reorganize the way food is produced, or starve.

James Howard Kunstler, *The Long Emergency*[1]

Arugula became an unlikely—and politically controversial—metonym for fresh produce and the escalating cost of food in America during the 2008 presidential campaign. While stumping in the Hawkeye State in the summer of 2007, Senator Barack Obama—then a front-runner for the Democratic nomination—lamented the rising cost of arugula sold at Whole Foods, portending that fresh produce and healthy food was fast becoming out of financial reach of middle and rural America. The irony of his statements is that they were made in Iowa, a state that does not have a single Whole Foods store, and that

Iowans, like most Americans, cannot afford much of Whole Foods' inventory anyway—arugula or otherwise. Iowa is also a state that produces an abundance of corn and soybeans, but not arugula. And Iowans, like many Americans, refer to arugula by its more common, English-derived name: rocket. Though pundits on both sides agree that Obama's "arugula moment" was a political gaffe, his underlying message was on point. Our nation's food—from beef, milk, and eggs, to corn, rice, and soy, and even to fresh produce like arugula—is, pardon the pun, *rocket*ing in price. And, as the price of oil has risen with the price of food and our economy has crumbled around us, Americans, for the first time in many generations, are beginning to understand what developing countries have always known: food security is economic security is national security.

In an open letter to the 2008 president-elect, journalist and best-selling author Michael Pollan outlined just how our current system of food production is compromising national security. Pollan argues that our complete reliance on fossil fuels for food production spells imminent catastrophe as the era of cheap and abundant—and nonrenewable—energy comes to a close. His arguments deftly illustrate the escalating futility of conventional agriculture. Pollan notes that in 1940, 1 calorie of fossil fuel energy produced 2.3 calories of food energy. But with today's industrial system of agriculture, the ratio has flipped to an inefficient, unsustainable equation, as it takes 10 calories of fossil-fuel energy to produce just 1 calorie of modern supermarket food. Pollan maintains that the solution "could not be simpler: we need to wean the American food system off its heavy 20th-century diet of fossil fuel and put it back on a diet of contemporary sunshine." He advocates for smaller agricultural efforts in more places across the country, "not as a matter of nostalgia for the agrarian past but as a matter of national security." Pollan further contends that "nations that lose the ability to substantially feed themselves will find themselves as gravely compromised in their international dealings as nations that depend on foreign sources of oil presently do. But while there are alternatives to oil, there are no alternatives to food."[2]

Pollan is not alone in his pessimistic views of our current state of food production. James Howard Kunstler, author of *The Long Emergency: Surviving the End of Oil, Climate Change, and Other Converging Catastrophes of the Twenty-*

First Century, is also a believer of the decimation that will ultimately result if we do not wean ourselves off of our high-petroleum diet. Many of Kunstler's arguments parallel Pollan's. Kunstler paints a chilling tale of doom for urban America that is quite frightening—frightening because his predictions do not seem particularly far-fetched. He predicts that smaller communities surrounded by agriculture have the highest hopes of surviving the Long Emergency. He is not so confident about the big cities, however, because they are growing in an unsustainable manner and they haven't had the urge to create or preserve an agricultural belt surrounding them. Kunstler concludes with a realization that our cities cannot continue to grow in the ways that they currently are, and predicts we will have to return to some form of agrarian life. No longer will large-scale industrial agriculture take place in entire states of Iowa, but that every city is going to have to be engaged in some form of food production.

Before we discount Kunstler's and Pollan's arguments as apocalyptic hyperbole, it is important to recall the many government-guided, community-implemented food production programs in this country that arose from national crises. The most significant—and prolific—of these were the Victory Gardens of World War II: Twenty million small gardens supplied 40 percent of the fresh vegetables consumed in America.[3] But there were similar food-producing efforts during the First World War, the Great Depression, and the Long Depression of the 1890s. During each of these distressed times, amid threats to national security, the federal government rallied the American people around food production, and created programs to educate citizens and assist them in exploiting food-growing opportunities throughout their urban communities.

The agriculture and gardening efforts during those periods of crisis were initiated to help secure our food supply, and the government looked to urban means of food production to supplement the rural farms that were unable to keep up with domestic demand. During World War I, the community agricultural efforts not only stabilized our nation's food supply, but bolstered that of the Allies as well. But more than a food source, the community agriculture efforts, especially the Victory Gardens of World War II, were meant to counteract a host of societal ills associated with crisis by providing "nutritional, psychological, and social returns for the individual and family."[4] These agricultural activities provided work relief for the unemployed; allowed the otherwise helpless women, children, and elderly to participate in the war efforts, giving them a sense of patriotic self-sacrifice; and even provided a form of recreation, allowing people to escape, if only momentarily, the troubles of the times.

Today, the need for similar public agriculture efforts could not be greater. In addition to the concerns that our earlier community food-producing efforts addressed, our current food system has far-reaching environmental and societal health ramifications as well. Principally, what is at stake is threefold: the rising cost of produce, and the resultant effect on our pocketbook; the degradation of our environment; and the growing girth of our citizens and the number of diseases associated with the obesity epidemic. The gardening and agriculture endeavors during our previous economic depressions and world wars helped supplement the nation's food supply and sustain the American population through periods of food shortages. The great irony today, however, is that the call for more abundant, locally led, and community organized forms of agriculture—even amid our current fiscal and war crises—is not so much an appeal to supplement our current system of food production, as it is to save us from it.

At the crux of both Pollan's and Kunstler's arguments is our nation's reliance on oil for the production of food. From before the advent of agriculture until the Industrial Revolution, societies never had to rely on fossil fuels to feed themselves. Today, the conventional system of agriculture in the United States relies on fossil fuels for almost every phase of food production: in the manufacturing of fertilizers, pesticides, and herbicides; for powering the complex machinery necessary for tilling, planting, harvesting, washing, sorting, and processing; and in transporting the final food product thousands of miles to our supermarkets. As the bounty of cheap oil dwindles, so too, does our bounty of food. The health of the people, and of our environment, will rely on restructuring how food is grown and delivered to the hundreds of millions of people concentrated in our urban environments. Smaller, localized agricultural efforts that do not rely on big, complex machinery, industrial agrichemicals, and vast systems of transport are needed in and around our cities. Fortunately, we already have an abundance of underutilized land within our cities—under public control—that can, at the very least, begin to return the agrarianism that Pollan and Kunstler contend is necessary for survival. Agrarianism and urbanism needn't be mutually exclusive.

Our carefree use of fossil fuels and their greenhouse gas emissions have also precipitated global warming and a dramatic change in climate, resulting in weather anomalies that are exploiting the vulnerabilities of our centralized agricultural system. Annual rainfall totals are diminishing in some areas of the country, and water is becoming increasingly scarce, particularly in the already moisture-deprived Southwest. As cities continue to sprawl into the desert, farmers are competing with urbanites for precious few resources. The Colorado River, the principal river of the region, is used so heavily to irrigate crops in California's Imperial Valley that it no longer consistently reaches the sea. Yet, water continues to be wasted on lavish fountain displays and verdant lawns throughout Southwest communities. As water scarcity increases with the rise of wasteful consumption and global climate change, it will be essential that purely ornamental landscapes be put to more productive use. Population centers like Las Vegas, Phoenix, and Albuquerque are going to have to figure out how to feed their citizens, and their only option may be to establish within their municipalities a local network of small farms or urban gardens utilizing dry-farming methods.

Water troubles are also plaguing the normally humid Southeast. That region's drought woes do not bode well for sprawling metropolises like Atlanta. Its dire water shortage became a topic of national dismay in 2007. Two years later, Atlanta is still grappling with a meager water supply. In 2008, the entire state of California also experienced parched conditions, threatening the state's abundant farms. The driest spring season in eighty-eight years left Governor Arnold Schwarzenegger little choice but to seek federal aid for California farmers.[5] During that same year, Iowa reeled from an overabundance of water. Much of the state experienced substantial flooding due to torrential rains, destroying more than 20 percent of the grain crop.[6] It is becoming abundantly clear that additional (call them "backup") systems of food production should be in place as we move into an uncertain climatic future. Coupled with the rise in oil and transportation costs, it is imperative that these systems be smaller, more abundant, and very close to home, within easy public access.

Our centralized system of agriculture is not only eroding our environment and economy, but our gustatory experience as well, erasing opportunities to enjoy fresh, fully ripened produce. Nonagenarian Juanita Kakalec reflects fondly on the times she used to pick fruit near her home in Washington, D.C. "It was just like milking a cow," she reminisced, recalling the simple pleasures of harvesting blueberries fresh from the bush, just a few miles north of the city, in Maryland. "You'd set your bucket down on the ground and just work your fingers over the branches, letting blueberries fall into the pail." Juanita also remembers picking strawberries, as well as visiting the peach and apple orchards in the area.

After her recent move to North Carolina, Juanita was looking forward to some local peaches. Though not as famous as their Georgian siblings further south, peaches grown in the Carolinas are wonderfully fragrant, juicy, and tasty. "Unfortunately, you can't find Carolina peaches here in the supermarkets of Carolina," lamented Juanita. "And when you do, they are not very good, because they pick them too early. It seems all the produce these days either comes from California or Peru."[7] Chile is more like it, but her point is valid.

Whether it is apples, avocados, or asparagus, the globalization of agriculture has given us year-round convenience. But when tied to the rising costs for oil, this convenience comes at a price. It raises the cost of produce and yields a diminished gustatory experience. It is a simple fact: Pickers have to harvest fruit before it is ripe so it can be shipped around the world without spoiling. Once the produce has been delivered, it is often gassed with ethylene to induce ripening. Global agriculture also favors cultivated varieties that pack tighter and bruise less, often sacrificing flavor and suppleness. The flavor, texture, aroma, and feel of a peach that is harvested early, transported thousands of miles, artificially ripened, then set on a supermarket shelf is quite different from one naturally ripened on the tree and plucked straight from the branch.

Juanita's desire for a fresh, local peach reminded me of an essay written by the provocative New Urbanist architect Daniel Solomon. Aptly titled *Peaches*, the essay relays the profound experiences fresh produce provide to the urban dweller. Solomon notes that "Food and urbanism are both fundamental to human experience." His argument is that the lack of everyday contact with fresh food in the modern city erodes our sense of place, disconnects us from the natural environment, and threatens an experience that was once commonplace. Solomon writes:

> Foodies worry that masses of people will go through life and never taste a peach that tastes like a peach. The people will survive somehow—it's *peachiness* that is threatened with extinction. In the contemporary world, retaining the full-blown potential of the flavor of a peach as a part of most people's life experience is no small matter. It involves land use policy, banking, union agreements, transportation, and distribution networks as much as it involves peach breeding, which itself is a more complex subject than ever before. In an agrarian society, where the peach trees are outside one's door, the perfect peach is commonplace. Delivering perfect peaches to the modern metropolis is another question.[8]

The land use-policies, transportation, and distribution networks that threaten our quest for perfect produce also threaten our pocketbook. Community food expert, Mark Winne, author of *Closing the Food Gap*, notes that the northeast region of the United States is especially susceptible. New England, at the extremity of both the national transportation system and the food chain, sees substantial increases in food costs compared to California, for example, where much of the country's fresh produce originates. As Winne contends, "The high energy costs associated with shipping food from those regions [near the beginning of the food chain] to New England increase food costs there by 6 to 10 percent."[9]

The distribution and transportation networks are not much shorter for communities in America's Heartland. According to *Food, Fuel, and Freeways*, a report by the Leopold Center for Sustainable Agriculture, the average produce item trucked to a terminal market in Chicago travels more than fifteen hundred miles. Grapes, broccoli, cauliflower, lettuce, green peas, and spinach all travel over two thousand miles to reach the Windy City. Most disheartening was the statistic for sweet corn. For Chicagoans, residents of the second-largest corn producing state in the nation, sweet corn travels, on average, 813 miles to reach them.[10]

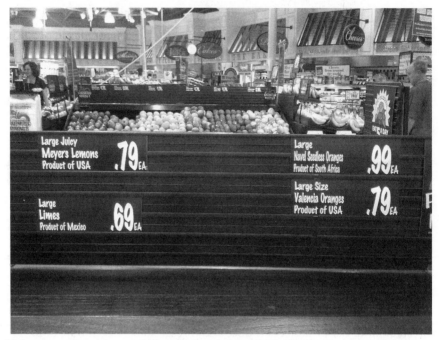

This supermarket citrus stand hints at our global system of agriculture, and the sheer distance that much of our produce has to travel to reach consumers.

As states have become more specialized in agricultural production, citizen access to locally available food has drastically diminished, erasing a bit of cultural heritage in the process. Take Iowa apples, for instance—a fruit with a long history in the Hawkeye State. The first recorded apple orchard in Iowa was planted in 1799, on the banks of the Mississippi River in Lee County.[11] By 1870, apple orchards flourished, and almost 100 percent of the apples consumed in the state were grown in Iowa. By 1925, apple production declined substantially, and Iowa produced just half of the apples consumed there. At the close of the twentieth century, apple production had all but disappeared: Only 15 percent of the apples consumed by Iowans were grown in their home state. Now, it is not just apples; almost all of Iowa's fresh produce supply is produced in other states and trucked in. It is estimated that less than 10 percent of the produce consumed in Iowa is grown in Iowa.[12] In 2007, fresh fruits, nuts, and vegetables represented just 0.13 percent of the state's cash receipts for all of Iowa's agricultural commodities (including livestock). Today, mink pelts produce three times the cash receipts as the state's apple crop.[13]

The specialization of conventional agriculture and its reliance on fossil fuels, coupled with global warming from greenhouse gas emissions and the resulting water scarcity and weather anomalies, are complex yet intertwined factors that contribute to the rising cost of food in this country. As if these were not reason enough for Americans to be wary of our current food production and distribution methods, there is yet another cause for concern: food safety. Foodborne illnesses resulting from pathogen-contaminated food is occurring with alarming regularity in this country, with the most widespread outbreaks happening in recent years. The pathogens most responsible for these food outbreaks are bacteria, specifically *E. coli* O157:H7 and various serotypes of *Salmonella*. Contaminations from these bacteria are typically associated with undercooked meat and eggs, though these pathogens are increasingly finding their way onto our fresh produce as well.

In response to the growing caseload of foodborne illnesses from fresh produce, the U.S. Food and Drug Administration (FDA) drafted the Produce Safety Action Plan. Initiated in 2004, the Action Plan outlines objectives and strategies to prevent contamination from pathogens and to minimize the public health impact when contamination occurs.[14] Even with the Action Plan in place, the FDA is finding it difficult to eradicate pathogen-infected produce and minimize the contamination's spread. Less than two years after the Produce Safety Action Plan was initiated, and in the face of recurring outbreaks of *E. coli* O157:H7 in fresh lettuce, the FDA drafted the Lettuce Safety Initiative. This initiative was aimed primarily at the California lettuce industry, and it sought to assess, document, and potentially regulate industry practices that demonstrated a risk of contaminating the lettuce crop.[15]

At the time the Lettuce Safety Initiative was published, *E. coli* O157:H7-contaminated spinach was beginning to infect people across the country. Over the course of two months, the pathogen-plagued produce sickened almost two hundred people from twenty-six states. Three people died from the outbreak. Two were elderly, and the other was a two-year-old child.[16] In the wake of yet another widespread and lethal microbial catastrophe, the FDA's Lettuce Safety

Initiative became the Leafy Greens Safety Initiative, and included a broader range of leafy vegetables, including spinach.[17]

In the summer of 2008, the Great Salsa Scare sent consumers of tomatoes and peppers into a panic when it was believed that *Salmonella* Saintpaul, previously considered a rare strain of the bacterium, was infecting people who had eaten fresh salsa (e.g., *pico de gallo*) from certain restaurants. At first, the outbreak seemed confined to Texas and New Mexico. But in the ensuing weeks, people across the country became sickened by *Salmonella* Saintpaul. When the number of reported infections finally ceased in August, four months after the first infections were documented, the *Salmonella*-tainted produce had sickened 1,442 people in forty-three states, the District of Columbia, and Canada. The Centers for Disease Control and Prevention (CDC) noted it was the largest outbreak of food-borne illness in the United States in the past decade.[18]

Not even a month had passed since that infamous record was set when people became sickened by another serotype of *Salmonella*. This time, *Salmonella* Typhimurium had contaminated peanuts processed by Peanut Corporation of America (PCA) at one of its plants in Blakely, Georgia. More than thirty million pounds of peanut products were recalled from stores, institutions, and even grade schools throughout the country, but not before significant health damage was wrought. From the first cases reported in September 2008 to April 2009, over seven hundred people were sickened across forty-six states. Nine of those people died.[19] Most unsettling, Stewart Parnell, the owner of PCA, knowingly distributed *Salmonella*-contaminated products. An FDA report submitted to the U.S. House of Representatives subcommittee investigating the outbreak noted that *Salmonella* was discovered in PCA's products a dozen different times dating back to June 2007. In 2006, an audit performed by Nestlé USA at PCA's Plainview, Texas, facility discovered fifty mouse carcasses in and around the plant, and a dead pigeon "lying on the ground near the peanut-receiving door."[20] The FDA report reveals that even though Parnell was notified by laboratories that his peanut products tested positive for *Salmonella*, he sold them anyway. "What is virtually unheard of," testified Charles Deibel, president of Deibel Laboratories Inc., one of the companies that tested PCA products for the Georgia facility, and found *Salmonella*, "is for an entity to disregard those results and place potentially contaminated products into the stream of commerce." Even up until January 2009, after the *Salmonella* Typhimurium outbreak was linked to peanut butter and peanut paste produced by PCA, Parnell pleaded with FDA officials that his workers "desperately at least need to turn the raw peanuts on our floor into money."[21]

In February of 2009, while illnesses from *Salmonella* Typhimurium were still being reported, *Salmonella* Saintpaul returned, this time contaminating alfalfa sprouts. As this book goes to press, more than two-hundred people across fourteen states have been sickened from eating tainted alfalfa, and the numbers have been rapidly escalating daily. The FDA and CDC are recommending that consumers not eat raw alfalfa sprouts until the agencies can trace the cause of the contamination. Though it looks like this outbreak will likely infect dozens more people, thankfully, no deaths have been reported.[22] Concurrently, pistachios processed in California have also tested positive for multiple serotypes of *Salmonella*, including Montevideo, Newport, and Senftenberg. As of May 2009, the CDC is still investigating whether this contamination has caused any human illness.[23]

With these frequent and frightening outbreaks comes an obvious uncertainty and general lack of confidence among Americans with regard to the security of our current food supply and distribution system. An Associated Press-Ipsos poll, conducted during the height of the *Salmonella* Saintpaul outbreak, found that almost half of adult Americans fear they may get sick from eating contaminated food. The uneasiness is more apparent with women and minorities. Only one in four women feel "very confident" about the safety of the food they buy. The most fearful group seems to be Hispanics. Half of the Hispanics polled had "little" or "no confidence" in the safety of the food they purchase.[24]

One strategy to help contain future outbreaks and boost consumer confidence is to require labeling that allows produce to be tracked from the dinner plate back to the farm, through the various retailers, processors, distributors, and packers. The lack of such a tracking system is why health officials in the country had a difficult time pinpointing the source of the *Salmonella* Saintpaul contamination. Early in the outbreak, the Food and Drug Administration believed the source of *Salmonella* Saintpaul to be raw red tomatoes, particularly plum, Roma, and round varieties. But officials never could find a definitive source of contaminated tomatoes. As the list of infected people grew, salsa was considered the common denominator, meaning that not only tomatoes, but cilantro and peppers became suspects as well. It took weeks before FDA investigators traced the source of contamination not to tomatoes, but to jalapeño peppers grown on a farm in Mexico. Later, serrano peppers from another farm in Mexico were implicated as well. Infected tomatoes are still believed to have been the source of the earlier *Salmonella* Saintpaul infections, but that hypothesis was never proven. The length of time of the investigation and the lack of definitive sources early on illuminates the vulnerability of this country's fruit and vegetable production and distribution.

Even if the necessary tracking measures are put in place, there is little to prevent contamination from occurring. As such, many people are still uneasy about food grown in distant parts of this country and in foreign countries. Confidence can only be guaranteed when there is complete transparency in the food system. It is not enough for some consumers to know where their food originates and how it got to the supermarket. Rather, these people demand to know—and to see—who is growing their food, where it is growing, and how it is being grown. Many want to talk to the farmers face to face, and even visit their fields and ask direct questions about pesticides and fertilizers. Meeting the people that grow the food you consume builds confidence and trust, and seems to be inherent with locally produced food. While the Associated Press-Ipsos poll was being conducted during the *Salmonella* Saintpaul outbreak, a reporter interviewed a grade-school teacher in Sacramento, California, about her thoughts on food safety. The teacher acknowledged that she buys most of her fresh produce from the local farmers' markets, and has largely resigned from supermarket produce. Her reasons are simple, "I see the same farmers every single week. You meet the people and you see where the [produce] is coming from."[25] It is this transparency in the food supply that gives people like this Central Valley schoolteacher comfort, and nothing could be more transparent than to have a source of food grown and harvested before your very eyes, as you travel from home to work, school, and places of worship, commerce, and recreation.

The recent outbreaks of pathogen-infected produce have certainly called into question the relative safety of conventional agriculture. But what about urban agriculture? There is a commonly shared perception that small, local farms and community garden plots produce better-tasting, healthier, and safer foods. But are they really safer than their factory-like agribusiness counterparts in remote regions of the world, or at least as safe? I posed the question to Marion Nestle, professor of Nutrition, Food Studies, and Public Health at New York University, and author of *Safe Foods: Bacteria, Biotechnology, and Bioterrorism.* Nestle, a proponent of urban agriculture who grows food on her twelfth-floor terrace in New York, acknowledged that the question regarding relative safety of urban- versus rural-grown food is hard to answer definitively. There would need to be testing of the specific produce items, she says; otherwise, we really do not have any way of knowing.[26] But the recent outbreaks of *E. coli* and *Salmonella* provide reasons to believe that there are perhaps inherent risks associated with our centralized system of agriculture that are simply not prevalent with local produce. A principal reason has to do with distribution. During the *E. coli*-contaminated spinach investigation, health officials determined the bacteria that sickened two hundred people in twenty-six states originated from one processor in San Juan Bautista, California. Likewise, the *Salmonella* Saintpaul that infected thousands of people across North America was traced to a single warehouse in Texas that received shipments from farms in Mexico. Ditto for the *Salmonella* Typhimurium-tainted peanut products, where over seven hundred people across forty-six states fell ill from the products of a single processor in Blakely, Georgia. In each instance, people across the country were sickened by eating produce or produce products originating from one locale.

"The bigger and more global the trade in food," Michael Pollan contends, "the more vulnerable the system is to catastrophe."[27] A decentralized system of many small, local farms and garden plots simply could never have the potential of infecting that many people over so large a geographic area. It was this very pattern of widespread infection that led health officials to conclude that, amid the thousands of people falling ill during the *Salmonella* Saintpaul outbreak, produce from local gardens was safe.[28]

Salmonella and *E. coli* are bacteria found in the intestines of animals and humans. So how do they get into our spinach, pepper, peanut, alfalfa, and other produce crops? Usually they come from the feces of animals, meaning that these bacteria can be found in the soil of our "pristine" farm fields (as is the case when fields are fertilized with manure), or, even more treacherous, in the water supply used to irrigate the crops. Indeed, irrigation water is a common source of microbial contamination of fresh produce. Large farm fields in the warmer and drier parts of the country (where most of our year-round fresh produce is derived) requires irrigation through large bodies of open water, such as canals and ponds. Open bodies of water present a potential health hazard, as they receive untreated stormwater runoff. When that stormwater finds its way into canals and ponds—after it has been in contact with chicken ranches, feed lots, cow pastures, and other places where concentrations of animal dung can be found— there exists a real risk of contamination. In fact, FDA officials traced the source of the *Salmonella* Saintpaul strain that infected serrano peppers to a holding pond used for irrigation.[29] Unlike their rural food-producing counterparts, urban agricultural efforts are at less risk from waterborne pathogens because they are not irrigated by large bodies of open water. Urban gardens are typically irrigated by rainfall and closed sources of potable water, delivered directly to the plants from city waterlines. The chance of contamination from water, therefore, is quite limited.

Of course, crops can also be infected from direct contact with fresh animal dung. This generally results in a more localized contamination, as the bacteria is not spread over an entire field as it is with manure or tainted irrigation water. But the risk is present nonetheless, and some have voiced concern about urban agriculture in public settings, and its proclivity to critters and their bacteria-laden waste. Animals are attracted to agriculture, certainly, but this condition exists in rural farm fields as much as it does on urban plots. In fact, rural fields typically have a greater abundance of livestock and wild critters than do our urban settings, and it is much more difficult to secure hundreds of acres of produce from animals than it is smaller garden plots in public spaces. Admittedly, where there are soil, water, and plants, there will be animals. Unless everything is grown in a secure environment—such as a greenhouse—it is virtually impossible to keep animals away.

Much of the concern regarding animals in urban settings is based largely on misperception. I find it interesting that we often perceive the suburban backyard vegetable garden or the rural farm field as pristine sources of fresh food, free from the harmful waste of animals and other critters. Yet, I recall the animals that are commonly found on rural farm fields: wayward livestock, coyotes, crows, gophers, mice, rabbits, snakes, and lizards. I also recall the animals that I found in my backyard gardens in California and Iowa: skunks, raccoons, deer, rabbits, squirrels, possums, the neighbor's cat (damn that cat!), and all sorts of birds, like cardinals, jays, robins, finches, sparrows, mourning doves, and hummingbirds. The birds, though their feces litter not only my garden but my patio furniture as well, are often welcome creatures to any garden—suburban, urban, or rural. The mammals are a bit more troublesome, for sure, but the fact is that animals will be present wherever food is grown, whether it is the perceived idyllic backyard vegetable garden, the fruit orchard in the neighborhood park, or the small family farm on the outskirts of town.

There may be reason to believe that certain animal waste in cities is safer than in farm fields, however. The CDC notes that reptiles, such as snakes and lizards, "are particularly likely to harbor *Salmonella*."[30] Reptiles are found in greater abundance on farm fields, but are less common in urban areas. As such, the risk from *Salmonella* poisoning could be much lower with produce grown in our generally reptile-free cities.

What about human waste? Another concern I occasionally hear is that the homeless will urinate all over the tomatoes and zucchini. Perhaps, although one hopes that homeless people realize that a system of public produce is the community's food supply, feeding not only the home owners and apartment dwellers in the area, but the homeless themselves. Maybe this is too idealistic. But if one searches for tips on sustainable gardening, and ways to improve the fertility of soil, one finds all sorts of strategies that make a stomach turn. It is well known that urine, high in urea (and thus, nitrogen), is a great fertilizer. Urine is also a good source of potassium and phosphorus as well, providing all three macronutrients that plants need. A quick Internet search yields thousands of articles espousing the virtues of human urine and gardening. Even more discomforting for the queasy, many women are using menstrual waste as an organic method of fertilizing their crops. Human feces, on the other hand, does pose considerable risks if used (or found) in the garden. Though there are a growing number of organizations promoting the benefits of "humanure" (the World Health Organization even published a paper on the topic, citing "the use of excreta and greywater in agriculture is increasingly considered a method combining water and nutrient recycling, increased household food security and improved nutrition for poor households"),[31] it is probably not suitable for smaller agriculture endeavors. The time frame for the breakdown of human manure is too long and the handling requirements too sophisticated for public-produce applications.

I realize this logic may not placate the fastidious, since we have become a society accustomed to produce with a shiny wax coating packaged on polystyrene trays and shrink-wrapped in cellophane. The bottom line is that all produce, whether purchased from the supermarket, the farmers' market, or grown in our backyard or a downtown parking lot, has undoubtedly come into contact with animals and insects, microbes and bacteria, and should therefore be thoroughly washed before consumption. And if there are particular public spaces with known problems of animal infestations or human encampments, then perhaps the best strategy is to seek another public plot.

Our industrial-scale, centralized system of food production is not only more susceptible to accidental contaminations from microbes, but from malicious terrorist tampering as well. Bioterrorism is a growing concern in this country, and for good reason. According to Marion Nestle, the demands placed on the Food and Drug Administration (which is tasked with monitoring the safety of the nation's fruit and vegetable supply and production, including imports), are unreasonable. Nestle reports, "About 700 FDA inspectors must oversee 30,000 food manufacturers and processors, 20,000 warehouses, 785,000 commercial and institutional food establishments, 128,000 grocery and convenience stores, and 1.5 million vending operations. The agency must also deal with food imports, which comprised 40% of the country's supply of fresh fruits and vegetables and 68% of the seafood in 2000."[32]

Because of the monumental burden placed on this severely understaffed government body, Nestle concludes that "It is not surprising that the FDA conducted only 5,000 inspections annually, visited less than 2% of the places under its jurisdiction, and inspected less than 1% of imported foods prior to 2001 when threats of bioterrorism forced improvements."[33]

The improvements Nestle alluded to have done little to boost confidence over our nation's food supply with regard to bioterrorism. During a press conference in 2004, after improvements to the FDA's funding and staffing were in place, Tommy Thompson, then secretary of Health and Human Services, offered a most chilling admission. Thompson told his audience, "I, for the life of me, cannot understand why the terrorists have not attacked our food supply, because it is so easy to do."[34] Michael Pollan agrees. "When a single factory is grinding 20 million hamburger patties in a week or washing 25 million servings of salad, a single terrorist armed with a canister of toxins can, at a stroke, poison millions." Pollan argues that "the best way to protect our food system against such threats is obvious: decentralize it."[35]

Another concern over the relative safety of conventional agriculture has to do with policing the use of myriad—and potentially harmful—agrichemicals. It is difficult to say what happens on those large agricultural fields in remote parts of the country (or in other countries, as China's melamine-contaminated milk scandal has proven) with regard to the application of chemical fertilizers, pesticides, herbicides, or other contaminants. The use of agrichemicals in conventional agriculture is for one purpose only: to increase profit by maximizing yields of saleable produce. As such, agrichemicals are often liberally sprayed on farm fields, and sometimes, on farm *workers*. One instance involving Ag-Mart (a prominent grower of tomatoes in North Carolina and Florida) and one of its field workers presents some of the grotesque effects that result from a person's direct exposure to agrichemicals. Carlos Herrera Candelario was born without arms or legs, and with abnormalities to his lungs and spine; the result of, according to his mother, repeated exposure to pesticides while she was pregnant.

Carlos's mother claimed she and other field workers were often doused with pesticides while they harvested tomatoes. Ag-Mart denies any wrongdoing, claiming the charges against the company are "a misreading of its records."[36] But between December 2004 and February 2005, three deformed children, Carlos included, were born to Ag-Mart field workers. Shortly afterward, Ag-Mart terminated its use of five pesticides that are known to cause birth defects. Without admitting guilt, Ag-Mart settled with Carlos's parents, agreeing to pay for his lifelong care.

Monetary profits are not generally the desired goal with urban public gardens, so maximum yields may neither be necessary nor desired. And people today generally desire organically grown produce. The problem is they cannot afford to buy organic, so they choose the cheaper, chemically grown produce. But with a system of public produce, where it is not financial gain that is sought, but community health, there is less reason to use chemical fertilizers, pesticides, and herbicides in the management of our urban food systems.

While urban soils may be free (or freer) of agrichemicals than rural fields, there is one contaminant that concerns many with regard to growing food in cities. Lead is often found in the grounds of our older city neighborhoods and former industrial areas. A common misperception, however, is that the presence of lead in soil automatically disqualifies any agricultural endeavor. Another misconception is that *all* urban soils are contaminated. There are generally just two sources for lead contamination in urban soils: lead-based paint, where peeling paint has fallen and mixed with the soil, and emissions from automobiles that ran on leaded gasoline. As such, the areas in the city where lead contamination may be likely are on the sites of old paint factories, gas stations, vacant lots where old buildings have been razed, near foundations of old buildings that may have been painted with lead-based paint, and within a couple feet of busy streets. While lead has typically been absent in paint and gasoline for quite some time, it moves little in the soil, creating a persistent concern for contamination.

According to a report by Carl Rosen, a soil scientist with the University of Minnesota Extension, "*The most serious source of exposure to soil lead is through direct ingestion (eating) of contaminated soil or dust* [emphasis his]. In general, plants do not absorb or accumulate lead." Rosen goes on to note that "Since plants do not take up large quantities of soil lead, the lead levels in soil considered safe for plants will be much higher than soil lead levels where eating of soil is a concern (pica). Generally, it has been considered safe to use garden produce grown in soils with total lead levels less than 300 ppm." At these levels and lower, the report states that "Studies have shown that lead does not readily accumulate in the fruiting parts of vegetable and fruit crops (e.g., corn, beans, squash, tomatoes, strawberries, apples)." Rosen states that leafy vegetables are more likely to absorb lead from the soil, but that there is "more concern about lead contamination from external lead on unwashed produce than from actual uptake by the plant itself."[37] The suggested remedy for external lead contamination? Wash your produce.

A simple site test is advised if the presence of lead is suspected. If lead is found, it may not be necessary to seek another plot to garden. As Rosen advises, contaminations of less than 300 parts per million are generally safe to garden without any soil remediation. If levels are higher, or if municipal officials want added peace of mind, there are numerous remedies to ensure safe, contaminant-free produce. The most common is to create raised beds on top of the existing soil. Remember, lead moves very little in the soil, and the risk of lead contaminating the upper soil is generally nil. Because lead tends to stay put, it is generally concentrated in the top three to four inches of existing soil. Another strategy is to excavate to this depth, replacing the contaminated soil with fresh, clean topsoil, which virtually guarantees contamination-free produce. A third strategy is to keep the soil pH neutral (i.e., 7.0), the level where the vast majority of plants thrive anyway. This can be done with common soil amendments. Soils with a pH of 6.5 or higher immobilize lead, rendering it unavailable to plants.

Other urban soil contaminants, such as paints, solvents, oil, gas, and other chemicals, are typically found in the same areas where lead can be common: gas stations, paint factories, and other former industrial sites. Municipal officials often desire to return these brownfields to green, and the remediation strategies for lead can be as successful as with other contaminants. Public space sites, however—such as parks, plazas, and town squares that were never previously developed, or subjected to chemical spills—are likely clean, requiring little remediation, if any. Still, a common soil test is recommended for any agriculture endeavor, rural or urban. If contaminants are found, and remediation proves too costly, sage advice is to simply find another site. The beauty of public-space cultivation is that there are many suitable—and clean—sites throughout the city.

There will always be risks associated with growing and consuming food. Some concerns are valid, though most are based on naïveté. Nevertheless, these perceptions may prove to be formidable obstacles to implementing a public-produce program in many communities. The truth is that farms today have few regulations in place to ensure absolute safety of fruits and vegetables, and there is perhaps greater potential for municipal government and its citizenry to work together to ensure a healthier—and safer—food system.

Colin Beavan, known by many as No Impact Man, set out on a year-long journey to find homeostasis, an equilibrium between his consumerist way of life and environmentalist ideals. His goal was deceptively modest: to sustain a simple life in New York City without making any net impact on the environment. To Beavan, that meant "no trash, no carbon emissions, no toxins in the water, no elevators, no subway, no products in packaging, no plastics, no air conditioning, no TV, no toilets. . . ."[38] And it also meant a very different way of eating. Beavan needed to eschew fast and processed foods, and only consume locally raised, organically grown foods to be honest to the No Impact Man project. At the end of his experiment, Beavan realized that "Eating local is a no-brainer if you live in a rich neighborhood with the cool, local-food farmers' market nearby." Beavan has been criticized that his experiment was bourgeois, and he now understands why. "Not consuming resources is no problem if a life of purchasing power has provided you with most of what you need," he admits. It is quite perplexing that to live a simple lifestyle in America is beyond the financial means of many. It is easy to say that we all should buy more organic, locally grown produce. It is quite another to be able to do so. And as Beavan has discovered, "Nutritious, local food should not just be available to the wealthy while the poor are left with McDonalds and KFC."[39]

Beavan's discovery of the conundrum between local, organically grown food and its high cost brings us to another important consideration in food security: public health. We've all seen the emaciated bodies of starving people living in countries crippled by food insecurity. It is incredibly oxymoronic that *obesity* is the result of food insecurity here in America. It is not the inaccessibility of food calories in this country that is problematic. Rather, it is the abundance of cheap calories derived from processed and fast food vis-à-vis the inaccessibility of fresh, wholesome, nutrient-dense foods at an affordable price that is responsible for the poor health of this nation's citizens.

We have found, through subsidizing grain crops and economies of scale, how to produce fast and processed foods in much larger quantities and cheaper prices than we can produce fresh fruits and vegetables. The bulk of corn produced in this country, for example, does not go to feed people directly. Rather, it is used primarily for silage to feed anything from cows (that produce meat, cheese, and milk), to chickens (meat and eggs) to hogs, and even to fish raised in fish farms. Corn is also processed into corn oil and high-fructose corn syrup, which has found its way into practically all of our baked goods, cereals, soft drinks, juice drinks, and other processed foods. In short, we have become a nation of corn.[40]

Corn—or more specifically, corn-derived food products—has now become the staple in the American diet. But cheeseburgers, soda pop, and snack foods have traditionally been regarded as luxury items, not staples; at least, they are not typically staples in those countries eating a traditional, non-Western diet. And certainly high-fructose corn syrup should be a luxury item or treat, as it is simply a sweetener. But through our subsidized and industrialized system of agriculture, we are able to produce these highly-processed luxury items so that they compete in price with fresh fruits and vegetables, nuts, seeds, pulses, legumes, and other grains; the types of food that should be staples in our diet.

Michael Pollan argues that "the surest way to escape the Western diet is simply to depart the realms it rules: the supermarket, the convenience store, and the fast-food outlet."[41] Instead, Pollan recommends eating more food from farmers' markets and community-supported agriculture groups. Easier said than done for some people. Consider the following: In July 2008, one dollar could buy a large, fresh, organic peach at the farmers' market, or it could purchase a double cheeseburger from McDonald's Dollar Menu. The peach has 73 calories and less than one gram of fat. The double cheeseburger has 440 calories, and

twenty-three grams of fat.[42] Which do you choose if you are hungry, impoverished, and living in a low-income neighborhood, and only have a dollar in your
pocket? It is really a trick question, as it is almost impossible to find fresh produce in economically depressed neighborhoods anyway. Fast food, on the other
hand, is ubiquitous. It is a harsh reality in a capitalist economy that supermarkets, farmers' markets, and grocery stores simply do not locate in impoverished
neighborhoods, leaving residents with a dearth of food options. Mark Winne
calls these areas *food deserts*—"places with too few choices of healthy and affordable food, and [that] are oversaturated with unhealthy food outlets such as
fast food joints." Winne explains that "while the failure of supermarkets to ade-
quately serve lower-income communities represents a failure of the marketplace, the marketplace is functioning rationally (as economists would say) by
going to where the money is." The consequential health outlook for people living in these food deserts is quite predictable. Residents of these areas, Winne
notes, "tend to be poorer and have fewer healthy food options, which in turn
contributes to their high overweight/obesity rates and diet-related illnesses such
as diabetes."[43]

Huntington, West Virginia, is one such food desert—perhaps the most barren in the nation. Once a proud and fairly prosperous coal-mining town, Huntington now carries the shameful moniker of the unhealthiest city in America,
according to statistics from the CDC. Almost fifty percent of adults in the Huntington metro region are obese. And that is just the beginning of the city's health
problems. Huntington leads the nation in heart disease, diabetes, and tooth de-
cay. Nearly half of all elderly adults in Huntington have lost *all* of their natural
teeth—an astounding statistic that no other city in the country can come close
to. A nurse at St. Mary's Regional Heart Institute in Huntington notes that many
patients are suffering from heart attacks in their thirties. At an age that is considered the prime of life in other parts of the country, people in Huntington are
getting open-heart surgeries. Hot dog eateries abound in Huntington. The city
has more pizza places than the entire state of West Virginia has health clubs and
gyms. "Fast food has become the staple," noted a manager within the state health
department, "with many residents convinced they can't afford to buy healthier
foods." A retired policeman blamed the economy, stating it needed to pick up

"so people can afford to get healthy." The city's mayor underwent stomach surgery to help him lose weight, yet he has no desire to curb the fast-food eateries that proliferate in Huntington. "We want as much business as we can have here," notes the mayor. "As many restaurants as you have, it kind of enhances the livability. Maybe not the health."[44]

On the other side of the country, municipal attitudes toward fast-food restaurants are considerably different. In the summer of 2008, the Los Angeles city council garnered national attention when it unanimously approved a one-year moratorium on fast-food restaurants within a particularly food-bleak section of their city. South Los Angeles is one of the more expansive food deserts in America, occupying thirty-two square miles and inhabited by half a million people. Like Huntington, the swelling of fast-food eateries in South L.A. is reflected in the community's expanding waistlines. This urban area has the highest concentration of fast-food eateries and the fewest number of grocery stores in the city. Thirty percent of South Los Angeles residents are obese, far greater than the 19 percent for the metropolitan region and 14 percent for the affluent area of Westside.[45] Residents of South Los Angeles also have the highest incidence of diabetes in Los Angeles County. To the city council, the need to suspend fast-food eateries is obvious. The health of their citizens is at stake, and the moratorium buys the municipality time to attract healthier food outlets.

As to be expected, restaurant associations and representatives of fast-food chains were dismayed, claiming the moratorium on fast food is misguided, and does not guarantee the emergence of healthier food options. And even if those healthier food options emerge, will they be affordable to the people of South Los Angeles? According to Kelly Brownell, director of the Rudd Center for Food Policy and Obesity at Yale University, people will change their diet when different foods are offered, but cost becomes an important factor in poor communities. Curtis English, a South Los Angeles resident who was interviewed by a reporter covering the moratorium, put the food problem in proper perspective. English recognizes that fast food is loaded with calories and cholesterol. But since he is unemployed and does not own a car, he is most concerned with how far he can stretch his food dollar within his neighborhood. English recalled that he ate at a McDonalds within a few blocks of his home twice the day before the city council passed the moratorium. For a mere $2.39, English had a sausage burrito for breakfast and a double cheeseburger for lunch. While Brownell notes that "Diets improve when healthy food establishments enter these neighborhoods,"[46] the real cost consideration is just how many healthful calories can one buy for $2.39?

A moratorium on fast-food establishments is a good start, but only solves one part of a more complex problem. As long as America is a capitalist nation, it is foolhardy to assume that supermarkets, farmers' markets, and restaurants with fresh, wholesome offerings will flock to distressed communities. Even if some pioneering establishments do choose to locate in a depressed area, will the healthy food offerings be affordable to the residents? The real solution boils down to accessibility *and* affordability. One strategy, and perhaps an effective one, is for the municipality to cultivate a policy that exploits the food growing and distribution potential of public spaces within these communities, to ensure that fresh, wholesome food is, at the very least, as prevalent as fast food, and just as cheap (or preferably, cheaper).

Though the fast-food moratorium is certainly controversial, the efforts of the City of Los Angeles should be lauded, as they illuminate the need for municipal planners and local government to tackle food insecurity in their communities head-on. Many communities across the nation have placed restrictions on fast-food restaurants, but they usually cite architectural design, or preservation of historical character as their reason. Los Angeles may be the only municipality in recent history to cite public health as the reason for its restriction. Though many object to having government interfere with private industry, the municipality's actions are really just an example of sound urban planning. A moratorium on fast-food businesses is no different from prohibiting a liquor store or an adult book store from locating near schools, for example, or requiring that manufacturing and heavy industry be segregated from housing. As David Zinczenko, editor in chief of *Men's Health* magazine and the author of several diet books, reasons, "What we're beginning to see is almost the monopolization of our dietary intake by a handful of corporations. Add to that the financial reality of feeding ourselves today, where a single grapefruit from a corner fruit stand costs two or more times as much as a few Chicken McNuggets, and I think you can begin to put together a case for governmental intervention."[47]

Los Angeles's moratorium on fast food demonstrates that municipality's belief that providing access to healthier food options falls well within the regimen of city planning and local public policy. At the very least, the council's actions open a dialogue about the specific roles city government can play to protect the community's health and welfare. Critics will continue to argue that the moratorium limits food choices, though the City of Los Angeles argues the contrary. The choice between fast food or no food is no choice at all. Los Angeles will, I predict, set a new trend in the planning and development of our cities, using food and public health as an organizer of city form.

It cannot be overstated: people living in dire conditions in this country need access to affordable, fresh, wholesome food in order to improve their health. Without regular access to affordable, nutrient-dense foods, our nation's waistline will continue to expand, and our health decline. The CDC reports that obesity rates across the American population have risen dramatically over the past three decades (a trend that coincides with the increase in availability of processed and fast foods). In 1990, not a single state in this country reported a prevalence of obesity greater than 15 percent of its adult population. In 2007, only one state (Colorado), had a prevalence of obesity less than 20 percent.[48] Among adults, obesity has doubled in this country over the past two decades. Today, one in three adults aged twenty years and older is obese in the United States. What is more alarming is the increase of obese children, from the very young to young adults. Data collected from two National Health and Nutrition Examination Surveys (1976–1980 and 2003–2006) illustrate this disconcerting trend. For children two- to five-years-old, the prevalence of obesity "increased from 5.0% to 12.4%; for those aged 6–11 years, prevalence increased from 6.5% to 17.0%; and for those aged 12–19 years, prevalence increased from 5.0% to 17.6%."[49]

The CDC has labeled American society "obesogenic," a condition resulting from "environments that promote increased food intake, nonhealthful foods, and physical inactivity."[50] Because we have created a culture inclined toward a

sedentary, overindulgent lifestyle, the CDC notes that the only way we can halt
obesity is through changes in policy and our environment. The CDC's Division
of Nutrition, Physical Activity and Obesity outlines six strategies to curb obesity, four of which focus on food. In addition to increasing physical activity and
decreasing television viewing, the CDC recommends that Americans decrease
the consumption of sugar-sweetened beverages; decrease the consumption of
high-energy-dense foods; increase breast-feeding initiation and duration for
newborns; and increase consumption of fruits and vegetables. Only 27 percent
of adults in America are eating the recommended three servings of vegetables
per day, and only 33 percent are meeting their daily recommendation of two
servings of fruit.[51]

The link between fresh produce and public health is so strong that even
health care organizations are devising strategies to increase accessibility to fresh
fruits and vegetables. Kaiser Permanente, one of the largest health-care organizations in the country, has recently instituted farmers' markets on the hospital
grounds of many of their facilities. Preston Maring, a Kaiser physician, came up
with the idea for a farmers' market after he noticed the success of the jewelry
and handbag vendors hawking their wares in the lobby of the Oakland hospital
where he practices. A firm believer of the connection between food, diet, and
health, Maring thought a modest produce stand or farmers' market could be an
amenity for patients and staff as well, perhaps even functioning as a form of pre-
ventative medicine. In 2003, the first Kaiser Permanente farmers' market
opened outside the lobby of the Oakland hospital. Two years later, two dozen
more opened in five states. Today, thirty farmers' markets operate in the parking
lots of Kaiser Permanente, from Georgia to Hawaii. What began as an idea by a
pioneering physician in Oakland, California, became a staple for Kaiser Permanente across the country, and a manifestation of Maring's belief that "Nothing is
more important to people's health than what they eat everyday."[52]

A bad diet affects not only physical health, but mental ability as well. According to a study published in the April 2008 edition of *Journal of School
Health*, students with an increased intake of fruits and vegetables fared better on
standardized literacy assessments than children on diets high in junk food.[53] For
this reason, and others regarding physical health, it is imperative that children

have access to a plentiful variety of fresh fruit and vegetables at home, at school, and on their way to and from these places.

Children are impressionable, and they tend to crave what they see around them. They are especially susceptible to the marketing blitzes of the big processed and fast-food companies. If children see nothing but ads promoting fast-food meals, they will want fast-food meals. A common ploy in supermarket chains is to place the sugary cereals, cookies, and other junk foods at eye level of children. This strategy might be tolerable if the marketing blitz were balanced with equally eye-popping graphics of fictional characters and personified animals touting healthy foods. Such is not the case. According to an article in the *New York Times*, "Almost three-fourths of the advertising aimed at children is for candy, snacks, sugary cereals or fast food."[54] Sweden bans all advertising aimed at children under twelve years old. Many other European countries restrict television ads during children's programming. But in the United States, marketing to "kid kustomers" is big business, as companies hope to snare brand loyalty at a young age, ensuring a customer for life.[55]

It is doubtful that Americans will pass legislation banning advertising to our kids anytime soon. Until then, healthy foods need to be just as visible and accessible as junk foods, preferably more so. Infusing our public spaces with fresh produce can help mitigate the marketing inundation of processed and fast foods, and actually teach children about the cycles of life, whole foods, and where those whole foods come from. If children really do crave what they see most often, ensuring the ubiquity of fresh produce is a strategy worthy of exploration.

Poor diet is not the only variable in obesity. Our sedentary lifestyle works to expand our waistline as well, and doctors routinely remind us that proper diet *and* exercise are the keys to healthful living. It is time to think how our public spaces could improve public health by providing places for exercise and access to healthy food. For example, the CDC states that one effective measure for combating obesity is to seek opportunities for physical activity within the community, such as hiking and biking along trails in parks and sidewalks along city streets. Not only could these public spaces provide opportunities for physical activity, but with the planting of fruits and vegetables, public space can increase access to the fresh produce that is necessary in (and largely missing from) American diets.

Such an example is already in place in Davenport, Iowa. Genesis Health System, a locally owned and operated health care facility for the Quad Cities, recently added a modern outdoor exercise station within the city's Duck Creek Parkway. In addition to the greenbelt's existing bicycle paths, playgrounds, and various sports fields and courts, the new fitness station offers another choice in physical activity. Genesis officials could not have erected their new exercise station in a more propitious location, near the shade of two very large apple trees. These remnants of what was likely a modest orchard provide the only clues to what existed here before the municipality purchased the farmstead and turned it into parkland. But those vestiges of local food production are strong. As the creek trickles by and cyclists silently pedal along its meandering path, people stair-step, push up, flex, and stretch, while red orbs of ripe fruit hang tantalizingly overhead. This active scene in such a serene setting sparks the desire for a healthier, more environmentally enriching lifestyle. Though there is a certain pastoral character to this particular park, the experience is uniquely urban. It is these experiences, rare today, that offer promise of a more bountiful, healthful, food-secure city.

If cities and their citizenry are to sustain, and realize enduring vigor and vitality, local systems of food production will have to be unearthed. As Michael Pollan notes, "The American people are paying more attention to food today than they have in decades, worrying not only about its price but about its safety, its provenance and its healthfulness. There is a gathering sense among the public that the industrial-food system is broken."[56] Pollan argues that until we address the flawed food system that feeds Americans, food security—and hence, national security—is compromised. James Howard Kunstler's claims are perhaps more dire. His apocalyptic forecast was easy to dismiss as a doomsday rant when *The Long Emergency* was published in 2005. But in the short time since, Kunstler's predictions are proving not only plausible, but imminent. For Kunstler, local food production in the twenty-first century is a simple issue of community existence: Those who produce their own food will continue to exist; those who cannot, will wither and die.

Though Pollan's and Kunstler's arguments tend toward hyperbole, their underlying message is grounded and quite lucid. The current agriculture system in America is proving vulnerable, and we need strategies to create a more secure food supply, for the health of our environment, our economy, and our people. Is large-scale agribusiness going away? Probably not. Is it reasonable for Americans to completely return to an agrarian lifestyle in and immediately near our cities? Doubtful. Is it possible to add produce choices and agricultural efforts in our intense urban settings, exploiting the food-producing potential of our current network of underutilized public spaces? Indubitably.

Successful public-space design in this country must respond to the needs and desires of a pluralistic society. The goal of the public-space designer is to ensure that the qualities and components comprising the physical space of the public realm provide the greatest value to all members of the community. When such requirements are met, public space becomes equitable, convivial, and communal. It is only natural that something as universal as the desire to eat healthy be fulfilled in our urban public spaces, and that these places teach us a thing or two about food, the environment, and each other. Nothing is more communal than diverse individuals coming together around food, and perhaps the time has come to consider public space as the community dinner table. Food is largely absent from public space today, but I do not believe there exists a more equitable approach to public-space design that provides the greatest value to its users, while building and strengthening community.

CHAPTER TWO

Public Space, Public Officials, Public Policy

I believe the city should own tracts of land for the growing of vegetables and fruits, where the citizens can see and understand that their real existence comes out of Mother Earth, and that the merchant or peddler is only a means of delivery.

Jens Jensen, *A Greater West Park System*[1]

In Davenport, Iowa, an edible garden is found in an unlikely place. Roma tomatoes, jalapeño peppers, variegated sage, and thyme are tucked among ornamental grasses and shrub roses in front of the municipality's Parking Office. This postage stamp-sized garden, barely a hundred square feet, punctuates one of the busiest corners in downtown. Countless people walk by the garden each day, with an occasional pedestrian grabbing a ripe tomato or a few sprigs of thyme. The parking manager, Tom Flaherty, enjoys his public garden, but wishes it had greater appeal to the passerby.

The following year Tom considers a trellis for training gourds, or some other flashy fruit that better announces the intent of his garden: "Public

Produce . . . Help Yourself." He decides pole beans are a good way to attract attention and a good food to offer the pedestrian. Other than pole beans, Tom plants marjoram, sage, rosemary, thyme, three varieties of peppers, and a cherry tomato vine. "I thought about going with the Romas again this year, but I think people were reluctant to pick them and carry them home. This way, people can just walk by, pluck a couple of cherry tomatoes and eat them right on the spot." He decides to plant the vine right behind a public bench, his logic being this location affords some protection to errant vandalism yet is close enough for people to reach their arm over the backrest if they want a quick snack.

People have asked Tom why he just doesn't place a sign that plainly states "Please eat our produce." "The Tragedy of the Commons," he cites, referring to the influential article written by the noted ecologist Garrett Hardin. Tom wants the public to harvest his produce, but stops short of a conspicuous message that

Punctuating one of the busiest street corners in downtown Davenport, Iowa, Thai peppers, pole beans, cherry tomatoes, jalapeño peppers, and various herbs are interspersed among shrub roses, ornamental grasses, and artwork outside the municipality's Parking Office. The produce is free for the taking.

encourages them to do so. He is concerned someone will harvest everything, leaving little for anybody else, and destroy the garden in the process. Instead, Tom prefers a more discreet approach as he believes more people will benefit. As he discovers, however, there is a delicate balance between a showy garden that subtly encourages the passerby to sample the fruit and a blatant invitation to do so. If the message is too subtle, pedestrians walk by without notice.

Tom harvests what the general public does not, giving much of the produce to coworkers and colleagues. On occasion, he sets up a table against a window in the Parking Office, in clear sight of passersby. He uses conspicuous signs on these occasions, inviting the public to help themselves to the various produce items. Supplementing items from his parking office garden with surplus from his home garden, Tom arranges vegetables and herbs on the makeshift produce stand. One week, Anaheims, Poblanos, and bell peppers are placed in a lid from a box of copy-machine paper with a note, "Come inside and help yourself to our

A makeshift produce stand inside the City of Davenport's Parking Office. This week, the offerings are local, organically grown herbs, courtesy of the Parking Manager's garden.

Parking Office Peppers." Another week, Tom brings in fresh herbs. He bags the different herbs and places sticky notes on each bag identifying the contents: Rosemary, Thyme, Cilantro, Oregano, and Fennel. His modest produce stand proves beneficial to a young man passing by. He stops in and asks if he could have some of the garden items. "Take as much as you like," an employee offers. The young man leaves momentarily and returns with a plastic shopping bag, stuffing it with handfuls of herbs.

The modest interest individuals have shown toward Tom's garden and ephemeral produce stands illustrate a clear need for more food choices downtown. Like many cities, Davenport has struggled for decades to attract a significant resident population back into its downtown. One of the greatest obstacles is the lack of everyday goods and services residents require—namely, food and other grocery items. Yet merchants and grocers are reluctant to locate downtown without a significant population base, creating a confounding circle of frustration. Interest is growing, fortunately, for both residents and merchants, and a few pioneers are staking claim. But the city is still far from providing a bounty of food choices in the downtown. In the interim, Tom tries to fill a need for fresh food, albeit with a very limited selection.

As the parking manager, Tom controls all the city-owned parking ramps and surface lots in the downtown, of which there are many—too many, actually, for the sparse resident and employee population in downtown Davenport. He struggles with one particular lot on the northern edge of downtown. Its area is half of an entire city block—about one acre—yet only a dozen or so cars park there on any given day. The city's planning staff view that parcel as a future development site. But it has been decades since that plot of land had any use more viable than a parking lot, and there are plenty of other properties in the downtown that will pique developer interest long before that acre of surface parking will.

Tom mulls over various strategies for the lot. He wants to maintain some of the area as parking for the few that use it, but what to do with the rest of the space? He decides on a half-acre community garden, hoping to create an amenity for the residents already living downtown, while possibly attracting more. If the resident interest proved low in the garden, Tom figured he could have his existing maintenance staff manage it as necessary, to keep it looking neat and tidy. After all, his staff already devotes a portion of their time maintaining the ornamental landscape around the other parking ramps and surface lots. Tom puts in a request to fund safety and aesthetic enhancements to the parking lot and construction of the garden through the city's Capital Improvement Program (CIP). Though unsure exactly what form the community garden will take, he believes it will meet broad community goals, such as reduce stormwater run-off, assist in neighborhood revitalization efforts, and even become a gateway into the downtown. His project is deemed worthy of funding, and his request is granted by the CIP committee: $370,000 of capital funds are allocated over the 2010 and 2011 fiscal years.

It is clear that Tom derives satisfaction from his gardens, both private and public. But his satisfaction is not that ordinarily afforded to gardeners. Aside from the typical physical and psychological therapy of working the land, getting one's hands dirty, and watching the miracle of nature turn a speck-sized seed into an edible bounty, the derivation of his satisfaction is different. His zeal for public gardening was cultivated once he became a government employee and realized just how much land is under public control. And over the years, he has realized that municipalities are often poor stewards of that land, not only with regard to the environment, but social equity, as well. "We have homeless people here—hungry people—and there has got be a way to extract more value from our public spaces," Tom exclaims. It is the equitable opportunities of public produce that give Tom his satisfaction, knowing that he can offer people free food. More to the point, Tom would like to give the underprivileged greater access to fresh produce. Tom enjoys growing tomatoes, for example, but doesn't particularly like the taste of them. But he understands the pleasure others derive from consuming vine-ripened, garden-fresh tomatoes, and it seems Tom enjoys giving his produce away as much as he enjoys growing it.

One day as Tom was harvesting Hungarian peppers, cherry tomatoes and green beans from his parking office garden, a middle-aged woman paused and asked what he was doing. Tom offered her some of the produce and she eagerly accepted. I asked if she was a street-person. "No," Tom replied, "but she is headed that way." He wished he could have talked with her more, but deemed she was not in a proper frame of mind at the time to carry a conversation. Nevertheless, Tom felt he made a positive difference in that woman's day and admitted, "that makes me feel good."

A block from the City of Davenport's Parking Office is the finest eating establishment in downtown: Duck City Bistro. Its owner, Chef Charles, contemplates growing vegetables in the adjoining Kaiserslautern Square, a handsomely designed civic space with a fountain, benches, enriched paving, and landscaping that, unfortunately, sees little social activity. It has been difficult in the past to keep the ornamental landscaping in this public square looking neat and tidy, and at times, the beds look a bit unkempt. Upset about the lack of maintenance and weeds growing in a raised bed alongside his bistro, Chef Charles came up with an idea to plant rosemary, basil, mint ("For *mojitos*," he added, "they are very popular"), leeks, and tomatoes for use in various menu items. The benefits of having a garden right next to the restaurant are many, not least of which is a more economical source of produce. Restaurateurs have to keep a careful watch on the rising cost of produce and the effect it has on their profit margins, which are quite slim, even for tony dining establishments. But Chef Charles noted that there were other benefits. "It would have meant more hours for my employees, because they would be the ones maintaining the garden." It was a point that I had never considered. Most restaurant employees work two jobs, simply because they cannot get enough hours during the work week to make ends meet. By allowing a garden adjacent to the restaurant, Chef Charles could have offered a few additional paid hours during the week to an employee or two.

The sight of a restaurant employee tending the garden, and harvesting produce that would be consumed by customers in a matter of moments, offered yet another benefit to Chef Charles. The restaurant business can be cutthroat, and restaurateurs are constantly seeking ways to distinguish themselves from the scores of other eateries in the city. A garden growing alongside the restaurant offers diners the most transparent form of food production. One can easily see not only where the food is grown, but what is growing, how it is maintained, and when it is harvested. Such transparency has tremendous appeal to the growing number of environmentally minded foodies who seek establishments that source food locally.

The benefits of maintaining a vegetable garden in the public plaza alongside Duck City Bistro were compelling enough for Chef Charles that he offered to enter into some kind of agreement with the municipality, such as a conditional use permit, whereby the city could stipulate that if his garden fell into a state of disrepair, the restaurateur could be fined and his gardening privileges revoked. But Parks and Recreation staff never responded to Charles's request. Assumedly, they did not want to give up that space, even though it has put an obvious strain on an already stretched maintenance crew. Instead, the Parks Department sent out laborers to rework the bed and plant shrub roses. Though the roses are quite attractive and add a nice touch to the plaza, their addition to the landscaped bed does not solve the ongoing maintenance burden placed on Parks staff, as Chef Charles has no desire to maintain something he cannot use in his restaurant. It seems that a win-win opportunity for both the restaurant and the Parks Department was missed.

What often comes to the minds of many when they hear "urban agriculture" is a community garden on a vacant parcel in a distressed neighborhood. The different gardening endeavors that are beginning to emerge in the small Midwestern city of Davenport, Iowa, hint at the variety of public spaces worthy of agricultural exploration, and the role that public officials could play toward implementing such endeavors. What is intriguing about these particular scenarios is that they all occur in the most intensely developed part of the city, where we find the greatest concentration of people and least expect to find agriculture: downtown. A downtown parking lot slated to become a community garden; a restaurateur looking to cut overhead costs while showcasing fresh produce in a downtown square; and a municipal official growing produce outside his office window, for anybody to harvest, on one of the busiest street corners in downtown—these individual efforts exemplify the creative strategies and places where urban agriculture has potential. The community garden in the distressed neighborhood will likely continue to epitomize urban agriculture. But as urban agriculture evolves in this country, it becomes clear that the diversity of public space within cities presents a diversity of food-growing opportunities.

I should clarify that by "public space" I am referring to those spaces that are freely accessible to the public, whether they are *truly* public or merely *perceived* to be. True public spaces include those properties owned and maintained by the municipality, such as streets and sidewalks, parks, squares and plazas, parking lots, and municipal buildings (libraries, city halls, and police and fire stations, for example, and the landscaped grounds that surround them). Civic institutions not owned by the municipality, but by other government or public agencies, may also be public, such as the grounds around courthouses, universities, and grade schools. Then there are those spaces that are privately owned, but where permission to pass is explicitly stated or implied. Hospitals, business parks, churches, corporate plazas, retail and commercial parking lots are examples of privately owned spaces where the public freely enters, and is often encouraged to do so. Even floodplains and transportation and utility easements, where structures are not allowed to be built, can be great opportunities for food production. In essence, any space where the public can enter throughout the day without being charged an admission fee (even if that space is privately owned and maintained), and that is suitable for growing food, is worthy of inclusion into a network of public produce.

I am not advocating the removal of fountains, benches, paving, sculpture, playground equipment, picnic tables, and other public-space amenities that attract people and make it accessible and comfortable for them for the sake of urban agriculture. Quite the contrary. I am interested in ways of attracting *more* people, by providing additional reasons for people to frequent public space, namely, wholesome, low-cost sustenance, food education, and a sense of self-sufficiency.

In the design of public spaces, there are many variables that, when properly identified and accommodated for, work together to create vivacity. Food is often one of those variables. This was something the late preeminent people-watcher William H. Whyte recognized almost thirty years ago. In his seminal book *The Social Life of Small Urban Spaces*, Whyte proffered, "If you want to seed a place with activity, put out food." That's because, he writes, "Food attracts people who attract more people." Whyte was so convinced of the positive impacts food has on the attractiveness of public space that he reiterated, just a couple paragraphs later, "Food, to repeat, draws people, and they draw more people."[2]

What Whyte was speaking about in particular was food prepared and sold from vendors, which helps make the many street corners and plazas in Manhattan so attractive to the passerby. Nevertheless, it is intriguing to ponder the effects that fresh, publicly accessible produce could have on the attractiveness of public space, and its ability to create a sense of conviviality. Other things being relatively equal, would the plaza with the orange grove or the apple trees be more compelling to people than one without any fruit? And what about produce vendors, like those on the street corners in Manhattan? What if a portion of their selection came from just down the street, in that pocket park or plaza? Would that be more enticing to the consumer?

Regardless if the space is truly public or only semipublic, municipal government is going to have to play a leading role. Programs, policies, funding strategies, and maintenance regimens of any urban agriculture endeavor will be difficult to implement and sustain if the largest land-owner in the city is indifferent. If public officials want a healthier, more prosperous citizenry, and believe that access to fresh, locally sourced, wholesome, and affordable food is good for both the individual citizen and the community at large, then public officials can no longer remain idle. In the face of rising food insecurity and declining public health stemming from a poor diet, public officials need to pursue various methods of providing better food choices in their community.

One of the easiest ways for municipal government to support a system of public produce is to simply *allow it*. Though attitudes are changing, most public agencies discourage or downright prohibit the planting of edibles in public spaces, largely over concerns about maintenance and perceived mess. (Such judgments are often based on misperceptions, which will be addressed in greater detail in Chapter 4.) These attitudes are especially prevalent with regard to streets, which is quite unfortunate. Streets represent the largest, most extensive network of public space in cities, and thus are significant places to explore edible landscaping, as every person in every neighborhood could be reached. Along many streets, there is a boulevard or planting strip between the sidewalk and curb. Some streets are even outfitted with wide, landscaped medians down their center. Historically regarded as aesthetic enhancements to streets, these landscaped areas are proving fundamental to the popular "Green Streets" movement, which is being implemented in cities like New York, Seattle, and Portland. Using landscaping to capture stormwater runoff, thereby reducing pollution of our lakes, streams, and rivers, green streets also help moderate air temperature, improve air quality, and provide habitat for urban wildlife. Boulevards and medians offer great potential for incorporating food-bearing plants in the streetscape, especially fruit- and nut-bearing trees and shrubs. These larger plants are not only desired to help define the street, and give neighborhoods character, but can more quickly and efficiently transpire larger amounts of stormwater runoff. Thus, incorporating agriculture along our streets helps communities attain broad equitable—and environmental—goals.

Looking closely at what gets planted in these public spaces, especially along residential streets, one will often find a pioneering homeowner extending his or her garden plot from the front yard to the curb. However, these residents are vigilantes, as almost every city in the nation prohibits fruit bearing trees in the public right-of-way. It is ironic that in places like San Francisco and Berkeley (municipalities that in many ways are leading the charge for better access to healthy, locally produced food), fruit- and nut-producing street trees are outlawed. San Francisco and Berkeley's urban forestry divisions operate in much the same way as other municipalities with regard to tree planting in the public right-of-way. If you want to plant a street tree, you need to obtain a permit. The City of San Francisco and the City of Berkeley do not plant fruit trees of any kind in the public right-of-way of their streets, and citizens wishing to plant a fruit tree in one of these strips will be denied a permit. Yet it happens anyway, on many streets throughout these two local-food-crazed communities, and municipal staff look the other way. But city government should take a more proactive role than turning a blind eye to enforcement of an unpopular ordinance.

The City of Portland, Oregon, is one municipality that does recognize the food-producing potential of city streets. Staff in that municipality's Parks and Recreation Department are seeking to codify the acceptance of fruit trees for their use as street trees. Though such a policy falls short of hearty encouragement to plant food-bearing trees in the public right-of-way, at least it absolves the owner of crime (or guilt) for wanting to establish some form of public food production. Even if public officials do not follow Portland's lead, and cannot be convinced of allowing fruit and nut trees along public streets, medians and boulevards still present excellent opportunities to plant smaller, tidier crops such as herbs and annual vegetables. In older urban neighborhoods, both front-yard and backyard spaces are modest. The strip of earth that separates the sidewalk and street allows residents to extend their garden plots. In some communities, this space is quite wide (five feet or more), providing an ample extension to the home garden. Street medians can be even wider, often measuring ten feet or

Top: In this generous space between the sidewalk and curb, a modest orchard has just been planted: Comice pear, apricot, Meyer lemon, and clementine surround an already established cherry tree. Of course, being across the street from the local farmers' market helps instill the desire for fresh, local produce. *Bottom:* Two houses down, the more typical, sterile landscape is seen: a monoculture of lawn that the owners obviously dread maintaining.

more in width. This generous size and their location in the middle of the street
give medians a more communal feel, and thus are great places to establish community gardens, which are easily accessible to everyone on the block.

Prohibitive language for food-bearing plants is not confined to street-tree
ordinances. The zoning ordinance for Davenport's central business district
(which municipal planners believe has never been updated since it was adopted
in the late 1930s or early 1940s), currently does not allow agriculture (commercial agriculture, specifically) anywhere in the downtown. Yet, these land uses are
among those that are permitted: creamery and dairy operations; flour, feed, and
grain (packaging, blending, and storing); commercial poultry and bird raising;
and fruit and vegetable processing. These point to a time when Davenport was
largely centered upon agricultural processing. But the *growing* of fresh produce,
at least for commercial purposes, was never accommodated in the ordinance. It
is probable that downtown agriculture would be permitted today if the intent
were to have gardens for people to freely forage. Because commercial agriculture
is not a permitted land use, however, the ordinance would likely disallow a
restaurateur like Chef Charles from growing herbs and vegetables that he then
sells to his dining patrons. And it would also deny downtown community gardening groups or small-scale entrepreneurial farmers the ability to sell their
produce at farmers' markets, a trend that is gaining in popularity across the
country as a way to promote local produce and help fund various urban agriculture efforts.

At first blush, it seems quite reasonable to discourage agriculture in the
downtown. After all, downtowns are for industry, retail, office, dense housing,
civic institutions, and traffic—lots and lots of traffic. Obviously, large-scale,
row-crop agriculture in the downtown is generally not an appropriate land use,
especially if it requires the demolition of buildings. But small-scale agriculture
ventures are sprouting throughout many urban areas, and downtowns make excellent places to extend such efforts. Amidst the office and residential towers,
there is a great diversity and concentration of both people and public space
downtown, making connections between folks and food easier. Squares, plazas,
pocket parks, and parking lots abound in downtowns, along with the grounds
around (and the roofs above) municipal and civic buildings. Collectively, these

spaces could provide a dense network of public produce accessible to the diversity of citizens commonly found downtown.

Of course, microclimatic conditions can be challenging downtown, with high-rise buildings casting deep shade over some areas throughout much of the day—not to mention the turbulent wind tunnels that are often experienced within the urban canyons of downtown. Downtown buildings also generate and reflect a lot of heat, creating a heat island that, while disadvantageous in some respects, especially with regard to energy use, could have beneficial impacts for urban agriculture for the northern climate cities in this country. Annual vegetables, for example, can often be started earlier—and extend later into the growing season—when planted downtown, because temperatures there tend to be warmer than in more remote, less-developed parts of the city. In general, the planting of gardens downtown should be encouraged, as gardens help improve air quality and moderate temperature, absorb stormwater, and provide much-needed greenery, softening the hard surfaces of the concrete canyons. The trick is to find those public spaces where even the most *micro* of microclimates is conducive to growing food. There are plenty. Even in the street-corner garden outside Davenport's Parking Office—a space surrounded by mid- and high-rise buildings and buffeted by wind—herbs, peppers, and tomatoes have done quite well (the pole beans, less so; lesson learned). While it may have made sense at one time to prohibit agriculture in downtowns, today small-scale agriculture can give downtowns a vital, life-sustaining vigor that is proving attractive for many urbanites.

Providence, Rhode Island, is not only looking at revising street-tree ordinances, or zoning restrictions in its downtown, but also working to incorporate permissible language for urban agriculture citywide. It is doing so through the strongest and broadest planning documents and policies available to any municipality. Community groups in Providence are working to double the amount of food grown in and around the city over the next ten years. That goal is unrealistic without sweeping revisions to current municipal codes. The Providence Urban Agriculture Policy Task Force recognizes that "If we are to redevelop and strengthen our local food web, agriculture at a variety of scales must be nestled into our region. In Providence this requires the calibration of planning and development policies to allow and promote appropriately scaled food production in diverse neighborhoods. Affecting this kind of change requires new language in Providence's Comprehensive Plan, followed by corresponding changes to the city's zoning ordinance, and new practices in affected city agencies."[3]

The City of Seattle's Comprehensive Plan could provide a good model for Providence. Seattle's planners consider urban agriculture an integral and necessary component of the city's network of managed open space. Within the "urban village" element of its comprehensive plan, the city adopted a goal of "one dedicated community garden for each 2,500 households in the Village with at least one dedicated garden site."[4]

The comprehensive plan is the most empowering document available to any municipality. The goals and policies set forth in a comprehensive plan help direct city ordinances, public policy, capital improvement projects, community programs, population growth, and development of land. Seattle's comprehensive plan demonstrates the municipality's commitment to urban agriculture, and offers hope of a new mindset for the increasing number of public officials across the country who believe growing food is not only an acceptable land use, but necessary for the health and well-being of the community and the environment.

Indifference among public officials is still pervasive, however, and is the most formidable obstacle to implementing any urban agriculture effort—whether on private or public land. Municipal staff in this country have especially been loathe to accept responsibility for feeding citizens directly. Once that attitude changes, and language is adopted to allow both private citizens and public officials to grow and maintain edibles, it will then be practical to devote time to exploring the myriad public-space opportunities for fresh, locally grown produce, and the roles public officials can play toward satiating the public's appetite for it.

Many municipalities would rather not repeal or rewrite restrictive zoning ordinances to appease persistent citizen groups seeking to improve their neighborhoods through urban agriculture. Instead, public officials merely grant permission to neighborhood groups to construct community gardens on abandoned lots in troubled neighborhoods. Indeed, vacant-land cultivation represents the standard—and future—of urban agriculture in the minds of many.

While the use of vacant lots to grow food can be an integral component of a successful network of public produce, vacant lots are not what we typically think of as public space, the sorts of places where concentrations of diverse people stroll through, or gather together to recreate, socialize, or simply pass the time. These lots are "public" merely because they have been abandoned, leaving the municipality with no choice but to assume ownership. In function, abandoned property is not public space—just simple open space. But this type of open-space cultivation does share some civic benefits with more traditional public space, namely, helping to build community.

The argument for vacant-lot cultivation is quite sensible: It allows land that nobody is interested in developing (at the time) to return to productive use, while lessening visual blight and bolstering community pride. Some community gardens have even helped to reduce crime in troubled neighborhoods, and have raised property values of adjacent structures. These obvious benefits give rise to an ironic new problem: By effectively creating the very situation that it sets out to establish, the successful community garden on a vacant lot increases the function and appeal of a dilapidated neighborhood, and that, in turn, increases development interest. In the minds of many public officials, community gardens on vacant lots serve only as placeholders until a developer is interested in improving the property. But to the community, the garden is a source of pride and good food, with years of sweat and toil poured into the soil. Raze the gardens to build homes, and you raise the frustration citizens have toward their government.

Such a situation is exactly what happened in New York during the late 1990s. New York City had a long list of active community gardens, some dating back to the early 1970s. Many in the community revered these green spaces, but to public officials, those garden sites were merely placeholders for future housing. In May 1998, Mayor Rudolph Giuliani transferred hundreds of community garden sites from the city's Parks Department to the Department of Housing Preservation and Development. This seemingly benign act spelled imminent doom for the gardens, as the opinions and public policies of Parks staff are very different from Housing and Development staff. To help generate revenue, the city's Office of Management and Budget mandated that the garden sites either be developed or auctioned. Mayor Giuliani's administration argued that the gardens were never meant to be permanent. The community argued otherwise, and a bitter green-bean war ensued. Protesters dressed as fruits and vegetables rallied outside the mayor's office, newspapers joined the fray and seemed to side with the gardeners, while Mayor Giuliani taunted, "Welcome to the era after communism."[5]

In the end, 113 garden sites were spared from development, but at a hefty cost. Trust for Public Land and New York Restoration, a community-based land trust led by entertainer Bette Midler, purchased the properties from the city for $4.2 million. Community groups declared victory, but, as one garden group noted, "forcing supporters of community gardens to pay the City millions of dollars to secure a future for community gardeners is bad public policy."[6] As New York and other municipalities across the country have learned, using community gardens as economic placeholders for future development is proving to be an unpopular strategy. If a garden site is successful, and has a group of dedicated citizens bent on improving the neighborhood and the lives of its inhabitants, it can be political suicide to try to take that land away.

What if community gardens were to make money? If urban agriculture was deemed a viable business to a plucky entrepreneur, would the city's stance on vacant-land cultivation change? University of Wisconsin professors Jerry Kaufman and Martin Bailkey sought to answer that question, and studied the extent to which entrepreneurial urban agriculture could be established on abandoned property in America. The impetus of their report is intriguing, as they cite the tens of thousands of vacant properties in each of the cities of Milwaukee, St. Louis, New Orleans, Chicago, Detroit, and Philadelphia that could reclaim productivity while establishing food security for these cities' food-poor citizens. The examples of the many urban agricultural efforts being attempted within various communities was heartening evidence of a nascent, national trend. In the end, however, Kaufman and Bailkey concluded their analysis with the realization that "City government leaders would like their middle-class residents to stay instead of moving to the suburbs. They wish for more market housing and small businesses located on vacant land. They would like to see a strong back-to-the-city movement to help fuel revitalization of depressed neighborhoods."[7] The pair could not find much support for their ideas even from venerable, far-sighted planners like Edmund Bacon. Their report recounted an argument Bacon made to the *Philadelphia Daily News* on his ninetieth birthday. Bacon urged planners and public officials to "wake up" to the amount of land that has been abandoned in their cities, and to find more rational uses for that land. Urban agriculture was not the rational use Bacon proffered, however. Instead, Bacon's strategy was to clear all vacant houses in order to assemble large tracts of obstruction-free land, which could entice suburban housing developers to build new neighborhoods.[8]

Kaufman and Bailkey reasonably argue that the middle class exodus continues in many American cities, and that considerable property—particularly that without the virtue of being near the city center or along a waterfront—will remain vacant and unsightly for the foreseeable future. Surely, in these areas, entrepreneurial urban agriculture makes planning sense. Such is the case in Detroit, a city estimated to have forty square miles of vacant land—30 percent of the city's total area. In the 1950s, Detroit's population was two million. Today, it is estimated to be less than half that, and still declining. Community groups have been turning many of these vacant parcels into food-growing opportunities for over a decade, and their efforts have inspired many. In some ways, Detroit is the embodiment of the National Vacant Properties Campaign slogan, "Creating opportunity from abandonment."[9] Now, thoughts are moving beyond the community garden plot to larger farming efforts. Some would like to see Detroit turn eyesore into opportunity by becoming the greenest city in the nation. Indeed, many urban planners see this bounty of empty land as a literal blank slate, with fantastic potential to reinvent Detroit. But even with all the vacant land, with more likely to come in the depressed economy, and amid projections that it would take at least an entire generation before Detroit could be repopulated, policy makers are still pining for the days when Detroit peaked at two million people. The thought of plants taking up space that could be inhabited by houses is a tough pill for some to swallow.[10]

Efforts to reinvent Detroit through urban agriculture give reason to believe that vacant-lot cultivation in other cities is a worthy revitalization strategy. However, many city planners and policy makers will likely continue to align with the planning strategy offered by Edmund Bacon. There is no denying the potential beauty and communal good that is possible with vacant-lot cultivation. But when there is an opportunity (or even hope) to bolster the tax base, create real density and diversity in the community, and revitalize a neighborhood with new homes and businesses, urban agriculture will seldom be seen as the highest and best use of abandoned land.

If a community insists on continued cultivation of vacant land in the face of a reluctant municipality, one option is to enlist the assistance of a land trust. Land trusts will acquire and hold land in perpetuity for the purpose of protecting that land from development.[11] There are many types of land trusts organized for many different purposes. Chicago's NeighborSpace, in particular, provides a unique land-trust model for municipally supported urban agriculture. NeighborSpace is a community-based, intergovernmental partnership between the City of Chicago, Chicago Parks District, and the Forest Preserve District of Cook County. Staff from each of these local agencies serve on NeighborSpace's board of directors, and each government partner commits $100,000 annually to acquire titles to vacant land, which they then deed to community groups who spare that land from development. There are many reasons to protect land from development, such as environmental conservation, historic preservation, land assembly for real estate speculation, and recreation, to name a few. What makes NeighborSpace unique is its pledge to "committed neighbors [who] have come together to turn vacant lots, railway, river embankments, and other open space into gardens and parks for community food production and beautification."[12] While many NeighborSpace sites are used for parks and ornamental gardens, protecting sites for the production of food is becoming more commonplace. It is this commitment to food production on urban open space and the active involvement and financial investment of local government officials that give promise to the tenure of urban agriculture on abandoned property.

Land trusts like NeighborSpace generally have excellent track records of successfully securing land for the preservation of open space, but there are times when an opponent proves too formidable. NeighborSpace was unsuccessful with one irregularly shaped, city-owned parcel on North Sheffield Avenue in Chicago, for example. It was not the municipality that objected to the proposal for an urban agriculture demonstration project, but the neighborhood. Residents overwhelmingly felt that the site's highest and best use was housing. In what was certainly a rare example of NIMBYism (a derivative of the acronym for "not in my backyard"), one that might provoke incredulity from many urban planners, neighbors argued that the community's appearance would be best improved not with green space, but with a building.[13]

There will always be controversy over what constitutes the highest and best use of abandoned property in struggling neighborhoods. During prolonged periods of economic woe, development declines sharply and hunger rises. Growing food for people is arguably the best use for land that lays fallow during such times. Indeed, it often takes such catastrophic collapses for public officials to reassess their public policies. While a few ardent activists have perennially advocated for better options in public transportation, for example, such pleas have historically fallen on deaf ears—until the price of gasoline leaped above four dollars per gallon.

Such is the case with urban agriculture. As we have moved into uncertain economic and climatic times, public officials across the country have taken notice of the nation's fragile food supply. The interest in growing food on vacant land to help establish food security has not been this strong since the Victory Garden efforts of World War II. But history since has taught us a lesson. That lesson is that the economy is cyclical, and we will witness prosperity again. And when those jubilant times come hence, the land that once laid fallow—that no one but gardeners and food growers would touch—becomes, once again, prime for development. As long as municipalities maintain control over vacant land, or uphold zoning regulations that restrict property to certain types of development, urban agriculture efforts on abandoned parcels will continue to be ephemeral. Only when the municipality relinquishes control of the land, or a long-term lease is agreed upon, will longevity be guaranteed to the community garden.

What could be a more permanent and acceptable strategy—to both citizens and public officials—is to look to other forms of public space in the city for urban agriculture. In any city, there are numerous under-utilized public and open spaces that could be used to produce food. According to Luc Mougeot, an expert on urban agriculture efforts around the world, "Municipal governments that have mapped their city's open spaces are amazed by how much space sits idle at any given time." He further contends that "Unused urban space is a wasted opportunity—an asset denied to a community's well-being and a brake on the city's development."[14]

Mougeot believes that urban agriculture strategies perform best when they can be retrofitted onto public and open spaces where other activities are already occurring. "Setting aside areas in or around the city for the exclusive and permanent use by urban agriculture is unrealistic and self-defeating," he argues. "For one thing, it ignores the economic reality of land prices in growing cities. More importantly, it misses out on the interactions that urban agriculture can have (and should have, if it is to prosper) with other urban activities." Instead, Mougeot urges municipal government to take a critical look at the myriad public and open spaces, and to ask probing questions, such as, "How much space in their city is unused, underused, or misused? Where? How much of this could be

This generous street median, which also serves as a utility easement, showcases a variety of drought-tolerant plants—native and naturalized, ornamental as well as edible.

made more attractive, more productive, and more profitable in social, economic, and environmental terms? How much could be achieved, in the short or
longer term, through urban agriculture?"[15] Public spaces that are too large for
the density of the surrounding development (suburban parks and parking lots);
too uncomfortable or uninteresting to attract a sufficient number of users
(many downtown plazas); or where development is neither capable nor allowed
(street rights-of-way, floodplains, utility and transportation easements) provide
great alternative sites to vacant lot cultivation. And these urban spaces, as they
are tucked in and around our places of employment, commerce, recreation, and
residence, provide that interaction of urban activities that Mougeot believes is
necessary in our cities.

The City of Des Moines is one municipality exploring that interaction between cultivation and community on its myriad public spaces. Through the
Community Gardening Coalition, a division within the Parks and Recreation
Department, municipal staff work with local landscape architects to create edible landscapes throughout the city. In addition to the community gardens that
have been established on the various institutional grounds of schoolyards, shelters, juvenile and family centers, food pantries, and libraries, true public space is
also being explored for urban agriculture. Drake Park, for example, has an orchard of thirty-three fruit trees at the south end of this neighborhood green
space, providing an ample supply of food for nearby residents. The plans for a
four-block section of East 12th Street, between I-235 and University Avenue, are
particularly innovative: grape arbors marking each end of this edible streetscape; fruit trees punctuating the street corners at each block; and raised beds
providing gardening space in the public right-of-way between the sidewalk and
curb. There is even a fruit and nut orchard planned along one block, creating an
attractive and edible edge to an otherwise unsightly parking lot. These edible
plants work in concert with other proposed streetscape amenities, like enriched
paving at street crossings, neighborhood signage, and sculpture set within a
traffic circle in the middle of one intersection. The aims of these neighborhood
plans are beautification and rejuvenation, but the designs go beyond the typical
visual enhancements. They help address issues of food security by providing a
diverse and abundant supply of fresh, publicly accessible produce.[16]

The City of Chicago is another municipality exploring local food production opportunities for the sake of its citizens' health and well-being. The city has adopted a pioneering food policy, dubbed "Eat Local Live Healthy," which identifies "food issues that, if restructured locally, could improve food quality, lower its cost and increase its availability for consumers."[17] The policy, authored by the municipality's Department of Planning and Development, identifies a framework of strategies that not only enhance public health, but creates food-related business opportunities and fosters public- and private-sector cooperation. Increasing food production in Chicago neighborhoods; improving access to locally grown, healthful food; and boosting public awareness of the availability and benefits of locally sourced food are just a few of the strategies outlined in Eat Local Live Healthy.

Cities like Chicago are ripe to take the next step in offering choices in locally sourced food on public land. And if the folks at City Hall are going to lead by example, then there is no better, more symbolic place to showcase public produce than City Hall itself. While visiting Germany in 2000, Mayor Richard M. Daley witnessed various aspects of urban agriculture, and was reportedly inspired to implement some of these efforts back home in the Windy City. An incredible opportunity for local food production was found right under his nose; or more specifically, over his head. Today, on the northwest corner of City Hall's roof, a colony of over 200,000 honey bees are churning out sweet rewards for this municipality's local food philosophy.

In 2003, shortly after construction was completed for City Hall's "green roof"—a garden in the sky that helps to insulate the building, reduce stormwater runoff, moderate air temperature, and provide habitat for butterflies and migratory birds—Daley asked two beekeepers from a local honey co-op to erect an apiary. Two hives of Italian honey bees were installed by Stephanie Averill and Michael Thompson, who manage the apiary and harvest its crop. The bees pollinate flowers as far as five miles from City Hall, returning with nectar to produce two seasonal—and two very distinct—blends of honey. During the spring and summer, the bulk of the nectar is collected from white clover, yielding a very light honey with superior taste. In the autumn, goldenrod and white aster nectar produce a darker and richer honey crop, better used for cooking. The honey, which is sold at the Chicago Cultural Center, the City of Chicago Store, and through the Internet is proving popular with locals and visitors alike. The proceeds from honey sales are funneled into the municipality's Department of Cultural Affairs to help support free public programs, such as art exhibits, performances, and other cultural events. Chicago's green roof provides a sterling example of the immense value that can be extracted from a typically forgotten public space. And the honey that is produced and sold proves that buying local food is not only good for the environment, but culture and community as well.[18]

In retrospect, Mayor Daley's directive to construct an apiary atop City Hall was prophetically visionary. Nations today—particularly the United States—are grappling with the startling decline in the honey bee population, referred to as Colony Collapse Disorder. And with the decline in honey bee populations comes a decline in food production. Many vegetable, fruit, and nut crops require pollination from honey bees. Human existence, we are quickly learning, is thus inextricably linked to these busy little bugs. Without bees to pollinate our plants, colony collapse disorder, as author Michael Shacker postulates, could well lead to "Civilization Collapse Disorder."[19] Treating honey bees as pests and controlling their population through insecticides is endangering the health of plants, the planet, and all of its inhabitants. Municipalities nationwide should be following Chicago's lead and erect apiaries to do their part to encourage active, productive honey bee colonies. While colony collapse disorder is still a mystery, and experts work feverishly to find its cause, the best we can do to end this syndrome is garden organically, and to take up beekeeping in our city spaces, so we can better understand—and appreciate—our wild pollinators.

The City of Chicago is unique in its top-down approach to issues of food security. Mayor Daley, with support from high-ranking public officials, is advocating for changes in food policy that have typically been lobbied for by grassroots groups. It seems the mayor and his staff truly understand the relationship between food security, community health, and economic prosperity, and are pioneering strategies to ensure that citizens have access to local, fresh food. In his introductory letter to Eat Local Live Healthy, Daley explains that "Local and fresh food would be most beneficial to our health, environment, and economy. But much of the produce we buy comes from places like California, Chile or

New Zealand. There are global environmental costs of shipping produce so far. And, the farther it is shipped, the less fresh it can be."[20]

What the City of Chicago succinctly illustrates is that public produce is an amalgam of public space, public officials, and public policy. If public produce is to be truly effective in bolstering the health and well-being of the city's citizens, municipalities must lead by example. Mayor Daley's views are not idealistic, in the pejorative sense, but they are no less visionary. His aims are high, but his expectations are reasonable. The mayor recognizes that commodity crops such as corn and soy bolster the city's economy, but they do not feed people directly. He warns Chicagoans of the need to restructure the city's food system to provide access to healthy, local table food. What Daley is sensibly advocating for are greater food choices in the community, choices that can improve both the health of his city's citizens and the health of his city's economy. The mayor readily admits that, given Chicago's northern climate, some food items will have to be imported. But in a plain-spoken manner that only Midwesterners have mastered, using the sort of pragmatic logic that is difficult to argue—and hence municipalities across the country would do well to adopt—Daley reasons that "Importing some food is different from importing most of it."[21]

CHAPTER THREE

To Glean and Forage in the City

Not everyone can afford to eat high-quality food in America, and that is shameful.

Michael Pollan, *In Defense of Food*[1]

At the end of the 2008 growing season, a farming couple outside of Denver, Colorado, opened their fields to anyone who wanted to gather potatoes, beets, carrots, and onions left over from the harvest. The Millers, owners of the farm, had never made such an offer before, but thought it could be a way to thank their customers while ensuring that perfectly good food did not go to waste. They arranged for the public giveaway to begin at 9:00 a.m. on the Saturday before Thanksgiving, and put the word out to the local media, thinking that over the course of the weekend, five thousand people might take them up on their offer. They underestimated. Forty thousand people—the size of a small city—arrived at the Miller farm to gather free food. People began lining up before dawn. By 8:30 a.m., the Millers' five-acre parking lot was full, and they had to direct cars out onto the open fields. The line of cars waiting to pick fresh, free produce extended over two miles down Highway 66. At the end of the first day of what was supposed to be a two-day harvest, the fields were picked clean, and an estimated 600,000 pounds of produce went home to grateful families.[2]

The public response to the Millers' benevolent gesture effectively demonstrates the hunger people have for fresh, free produce. But there is more to unearth from this example than just people making a trek for free food. It was made clear to anybody wanting food from the Millers' farm that they would have to pick it from the fields themselves. I suspect if the Millers had instead promised to hand out sacks of onions and potatoes to anybody who showed up, the response would not have been as large. There can be a certain shame that accompanies the acceptance of handouts, as it is often taken as a sign of poverty. Working for one's meals is certainly more respectable, and it may have been the opportunity for the public to labor for its rewards, rather than merely accepting charity, that provoked the large response. More to the point, it was the prospect of harvesting, the specific act of venturing out into the fields and pulling food from the earth—proof to one's self that he or she has the wherewithal to provide for the family—that provided as much compulsion for folks as the food's gratis price tag, perhaps more so. The huge crowd that turned out to pick produce reflected the desire people have for a bit of agrarianism in their urban lifestyles.

The practice of gleaning—gathering food left over in the fields from the commercial harvest—is an age-old manner of putting food on the table. In many parts of the world, though, gleaning has long been stigmatized. While some argue that it is entirely appropriate (and perfectly respectable) to gather food that would otherwise rot in the fields, others consider it a pitiful endeavor, as it is usually undertaken only by those less fortunate. The French realist painter Jean-François Millet sought to erase the stigma of gathering food left behind by others in his brilliant mid-nineteenth century portrait *Les Glaneuses* (*The Gleaners*). Millet depicts a melancholy scene: peasant women, in the twilight of the day, their backs hunched over the harvested fields of rural France, gathering leftover grains that lay on the ground with their dirty, masculine hands. Yet, Millet expertly portrays an air of dignity within the scene. The subjects are gleaners, after all, not beggars, and there is pride in an honest day's toil.

Jean François-Millet's *Les Glaneuses* (The Gleaners), 1857.

These women, though they stoop for scraps, are not to be pitied. They ably provide for themselves and their families.

Gleaning continues in France to this day, and a provocative film, entitled *The Gleaners and I*, documents how this practice has evolved over the last 150 years. Building on the dignity depicted in Millet's portrait, producer Agnès Varda poignantly illustrates how gleaning continues to be a respectable method of providing for oneself and one's family. As revealed in the film, released at the turn of this century, women are no longer the sole gleaners, as was depicted in Millet's portrait and in others painted at the time. Men and children are now commonly seen gleaning the fields of France.[3]

Varda focuses on the myriad groups that still benefit from gleaning—from today's rural peasants to urban artists. One of the more intriguing beneficiary groups is restaurateurs. One subject, a gourmet chef, gleans simply because of the steep overhead of the restaurant business, and survival requires frugality. The chef, the youngest in France to have received two stars from the esteemed

Michelin Guide, laments the huge price that savory and other fresh herbs command. To combat the high price of quality ingredients, he roams the nearby fields in the morning, picking and gleaning what is necessary for the day's menu. Not only does this save the restaurant money, but gives the chef an assurance of quality: He knows exactly from where and when the produce was harvested. For this same reason, other subjects in Varda's film glean hundreds of pounds of potatoes, for example, to sell to other restaurants that are looking for fresh, locally grown produce at an attractive price. Some of these potatoes are too ugly and misshapen to be saleable in any market. Others are simply too large. Only those potatoes that are of a certain caliber, and are free from blemishes, are delivered to market; all others—about twenty-five tons—are dumped in the fields to rot because consumers supposedly will not waste good money on big or ugly produce. Restaurateurs, however, are always looking for ways to trim the costs of quality ingredients, and they buy the gleaned potatoes by the bushels. It does not matter what the potatoes look like to the restaurateur, because after he or she is done with them, nobody will know that they were once ugly. For chefs, what matters most are flavor, texture, and freshness, and even though these potatoes at one time looked unpalatable to the markets, they are just as flavorful and texturally rich as their more handsome siblings.

The Gleaners and I reveals that gleaning is still a practice undertaken—and even enjoyed—by many in France. But what about here at home? Certainly, gleaning is practiced throughout America's urban areas as well. Unlike Millet and Varda's subjects, however, America's gleaners do deserve our pity. The needy gather scraps of food from the trash bins of bakeries, cafés and restaurants—food that may contain harmful bacteria such as *Salmonella* or *E. coli*. They pick fruits and vegetables off the ground after the farmers' markets have closed; most of what they find has been trampled upon or was unfit for sale. They roam the streets of our cities, gathering cans and bottles from trash bins. They pull shopping carts great distances to the local recycling center, only to receive a relatively paltry reward for their effort. Gleaning from our urban environment, it seems, is a far cry from the much more healthful and bountiful yields in rural France.

What is needed are healthier opportunities to glean fresh produce, similar to those demonstrated on the Millers' farm, only closer to home. That means the ability to glean on farm fields within walking or biking distance, or a transit ride, from one's kitchen. But the establishment of larger-scaled agriculture in some cities may be difficult, if not impossible, given the intensity of existing development and subsequent land constraints—or, more pervasive, the desire among public officials to develop existing farmland into housing. While it is imprudent to raze homes for the sake of urban agriculture, it is possible—and good practice—to mitigate overzealous greenfield development (i.e., constructing buildings and streets on otherwise pristine, undeveloped property) by preserving existing agricultural land, and incorporating opportunities to farm into new subdivisions and towns. Even those who are not professional planners recognize the need to place restrictions on greenfield development, and to treat fertile farmland with the same preservationist ethic as we do our most prized open spaces and natural resources. Knight Professor of Journalism *cum* food-security expert Michael Pollan, of the University of California, Berkeley, offers a couple of strategies for municipal officials:

> In the same way that when we came to recognize the supreme ecological value of wetlands we erected high bars to their development, we need to recognize the value of farmland to our national security and require real-estate developers to do "food-system impact statements" before development begins. We should also create tax and zoning incentives for developers to incorporate farmland (as they now do "open space") in their subdivision plans; all those subdivisions now ringing golf courses could someday have diversified farms at their center.[4]

Andres Duany's actions echo Pollan's proposals. During a forum in Charlotte, North Carolina, Duany, the renowned New Urbanist architect, called agriculture "the new golf," and quipped "Food is quite good-looking," meaning views of farmland can be just as attractive to home buyers as views of a golf course.[5] Duany noted that developers across America have been asking him to capitalize on the growing desire for locally grown food by incorporating farmland into new subdivisions. New Town at St. Charles, a 726-acre suburban development thirty miles northwest of St. Louis, designed by Duany and his firm, explores such an integration between farm and community. Unlike conventional greenfield subdivisions, which are comprised entirely of single-family detached homes, New Town's designers integrated a variety of public spaces, goods, and services into the development. What is the most forward thinking component of the development, however, was the incorporation of a complete turnkey farmstead, with the goal that organic produce raised can be sold directly to the community. And if anything is left over from the harvest, then the potential to glean exists as well. The attraction to gleaning may be quite high for New Town residents, as the farm is not located in the hinterlands beyond the development, but is within walking distance of many homes and public spaces. The more convenient it is to gather fresh produce, the more likely people will do so.

Of course, the intention is for all of the produce raised to be sold back to the community. But with most farms, gleaning is possible because there is almost always produce left in the fields from the harvest. Neither harvesting crews nor machines are 100 percent efficient, and as the harvest progresses from row to row, there are usually grains, tubers, berries, and leaves missed by the machines or the hands of harvesters. Those in the community that can afford to buy freshly harvested, locally grown produce can now do so. And for those who cannot, or who simply revel in the prospect of a bit of agrarianism, the farmstead at New Town could offer the opportunity to glean.

New Town at St. Charles is not without criticism, however. In the dubious tradition of suburban sprawl, New Town is a greenfield development through and through, developed beyond the urbanized edge of Saint Charles, on hundreds of acres of fertile farmland. Though a portion of that soil was preserved within the development, over 700 acres were obliterated for streets, buildings, lakes, and canals. Nevertheless, this suburban development is a vast improvement over the traditional greenfield developments that typically comprise little more than single-family homes in a sterile landscape. This development maintains some level of agricultural productivity, and the farm is a highly desirable amenity and food source for the residents. Planners, urban designers, and even developers, are recognizing "farming as another mixed use that adds vitality to the community."[6] After a half-century hiatus, agriculture is finding a place next to public spaces, entertainment and cultural venues, employment opportunities, recreational amenities, educational institutions, and shops and markets as principal constituents of great communities.

While urban farms provide a marvelous opportunity for gleaners, there are a few obstacles to overcome. For one, gleaning requires mutual consent. In other words, gleaning is typically carried out on private land, and thus requires the willful participation of the landowner. Some, like the Millers, may be willing to open their fields to the public; others may be frightened by the thought of thousands of people descending onto their property. Even if there was a pervasive culture of sharing, space constraints within cities can limit opportunities for larger-scaled farming, and thus gleaning. Also, gleaning takes place only after the commercial harvest is over, presenting a more limited supply and shorter window of time to gather food than if fields were open to anyone throughout the growing season. For these reasons, and others, gleaning alone cannot satisfy the urbanite's insatiable hunger for fresh, free produce. There need to be opportunities to truly forage in the city as well.

In his revelatory book *The Omnivore's Dilemma*, Pollan sought to better answer the age-old question, "What should we have for dinner?" After traipsing through the cornfields of Iowa, chicken ranches in Virginia, and feedlots, food science laboratories, and organic mega-ranches throughout the country, he finds himself back at home in the Bay Area, on a quest to find dessert. Pollan recollects:

> My plan was to forage fruit, for a tart, from one of the many fruit trees lining the streets in Berkeley. I see no reason why foraging for food should be restricted to the countryside, so . . . I embarked on several urban scouting expeditions in quest of dessert. Actually, these were just strolls around the neighborhood with a baggie [sic]. In the two years we've lived in Berkeley I've located a handful of excellent fruit trees—plum, apple, apricot, and fig—offering publicly accessible branches.[7]

Having lived in Berkeley for many years myself, I can attest to the bounty of food-producing shrubs and trees that line the neighborhood streets. In addition to those that Pollan had found, I have seen oranges, lemons, cherries, persimmon, fennel, the occasional tomato vine, as well as rosemary, thyme, sage, and other herbs—all occupying space between the sidewalk and the street. Though these plants may have been planted and tended by private residents, their location in the public right-of-way means they now belong to everybody. In short, it *is* entirely plausible to find dessert—and more—merely by strolling the streets of Berkeley.

Pollan's quest to forage in the city was more than a conceit. While he admits gathering entire meals from one's urban surrounds is probably not feasible on a regular basis, he believes it is an important endeavor undertaken occasionally to remind us where our food comes from. Much of the food security problems that are facing the nation today stem from this lack of knowledge, Pollan argues. Understanding where and how food is made available, and at what time, offers many benefits, not the least of which is the personal challenge to provide for oneself and succeed, providing satiety for the both the body and spirit.

Surely one can learn about the nature and culture of human eating, as Pollan hoped to do foraging for ingredients for his dessert. But gathering food from the urban environment can also yield more tangible benefits. Not only is there greater environmental responsibility in gathering fresh produce from the source, rather than having it trucked in from a distant region, but gathering food can supplement caloric intake by providing nourishment in the form of snacks, or maybe even complete meals. For children, being able to forage for fresh produce in the city may mean the difference between a bag of chips on the way home from school or an apple. For the working poor, it may mean the difference between skipping meals in order to pay the electric bill or a healthy dinner. For the utterly destitute, it may mean the difference between foraging for fruit in a public plaza or for scraps in a dumpster. Foraging exploits the social equity potential of public produce for those hit hardest by the rising cost of fresh food, or those who do not have ready access to it.

In Los Angeles, a grassroots organization known as Fallen Fruit promotes social equity, public health, and environmental stewardship through the act of foraging for public fruit. The group, founded by three Los Angeles artists, un-earthed an arcane city ordinance—a usufruct law—that states that fruits over-hanging any public space, regardless if the tree is planted on private property or not, are public goods. In general usage, usufruct laws give a person legal access to somebody else's property, provided that the property is not damaged. What this means in the city of Los Angeles is that fruits that can be plucked from city sidewalks, parks, and even semipublic spaces like parking lots and plazas where permission to pass is granted to the public, are considered fair game, and pro-tected by law. Fallen Fruit believes that "fruit is a resource that should be com-monly shared, like shells from the beach or mushrooms from the forest. Our goal is to get people thinking about the life and vitality of our neighborhoods and to consider how we can change the dynamic of our cities and common val-ues."[8] Artwork on Fallen Fruit's website is captioned with catchy phrases, such as "Eat Local, Think Global," and "Public Fruit = Public Health." One image poses the question, "Why live with barren landscapes and sterile shrubs?" recognizing that our urban environments could be much more livable with a simple land-scape conversion, from sterility to fertility. The Fallen Fruit mission is certainly inspiring, and the organization's actions profoundly parallel the fundamental arguments in this book. In an effort to achieve these lofty goals through forag-ing for food, the group publishes neighborhood maps via its website illustrating where usufruct fruit can be found and harvested. The efforts of Fallen Fruit are making worldwide news, and it seems that the organization's ideas are not only spreading roots throughout Los Angeles, but in other communities as well. At last check, for example, the Second Street neighborhood in Santa Fe, New Mex-ico, had been included on the Fallen Fruit website, with map locations shown for chokecherry, mulberry, plum, piñon pine (pine nuts), apricot, golden cur-rants, and other fruits that thrive in that high-desert area of the Southwest. There is even a legend denoting when the various fruits ripen: plums and apri-cots in the summer, grapes and apples in the late summer or fall, for instance. In addition to mapping public fruit, Fallen Fruit now helps plan "fruit parks" in underutilized public spaces, providing more opportunity to augment a commu-nity's supply of fresh produce.

The positive public response to both Michael Pollan's and Fallen Fruit's ideals demonstrate the desire by many for better food education, environmental responsibility, and perhaps most importantly, social equity through better food choices. Foraging—like gleaning—has also been a cost effective way for individuals and families to put nutritious food on the table, and the ability to harvest public produce presents a great opportunity for those facing hunger or who generally can only afford to eat poorly. The inconvenient truth in the Land of Plenty is that families nationwide do not have access to food that is both nutritious *and* affordable. This became glaringly apparent in the 1980s with Hands Across America, a benefit to raise money to help fight poverty and hunger. During the campaign, communities came to recognize the sheer numbers of hungry individuals and families locally and throughout the country. In Hartford, Connecticut, for example, a study conducted by the Community Childhood Hunger Identification Project (CCHIP) found that up to forty thousand residents were struggling to feed themselves or members of their families. Interestingly, the CCHIP did not recommend more emergency food programs (as they were deemed feckless). Instead, the study prescribed increasing accessibility to fresh, wholesome, affordable food. Some methods listed were farmers' markets and food stores within or at least closer to impoverished neighborhoods, but the study also advocated for more community gardening and other methods where fresh produce could be close at hand with little to no cost to the consumer.[9]

American cities today are grappling with a shrinking middle class and a growing number of have-nots, particularly in the wake of our current recession. Hunger is usually—and erroneously—associated with the down-and-out, such as the homeless. The truth about hunger is more obvious today, as its reach extends to the millions of people who have lost their jobs during the downward spiral of our economy. Families nationwide are now cutting the size of their meals—or cutting out meals entirely—to make ends meet, endangering their health and, most alarming, the development of their children. As Gleaners, a food bank serving central Indiana, reminds us, "America's hungry are more than 'street people' you see carrying torn trash bags bulging with aluminum cans or pushing shopping carts half filled with somebody's discarded clothing. This country's hungry are also the elderly . . . the working poor . . . single parent households . . . and children, so very many children."[10]

Even families in areas where the cost of living is high, families whose incomes would otherwise put them squarely in the middle class in other areas of the country, are having to forego meals to pay bills. In San Francisco, one of America's most expensive cities in which to reside, the number of people facing hunger is staggering. According to the San Francisco Food Bank, some 150,000 San Franciscans in the 750,000-person city—20 percent of the population—forego food (i.e., skipping dinner, eating less, or eating less well) in order to pay the rent. The situation is direr for children and the elderly. The same food organization reports that "1 in 4 children and 1 in 4 seniors do not have access to enough food to meet their nutritional needs on a regular basis." The irony is that many of the individuals and families that need food assistance in San Francisco have salaries well above the income cut-off for Food Stamps and other government nutrition programs. As such, hunger is not limited to one or two impoverished areas of this high-rent city. "In most neighborhoods, a *minimum* of 10% of the population experiences some kind of disruption to their daily nutritional needs."[11] Today, the hungry are even more prevalent. Since the economic downturn and the alarming rise in joblessness, food banks nationwide have seen anywhere from a 15 to 30 percent increase in the number of people demanding food assistance.[12] An opportunity to forage within each and every neighborhood can help stave off hunger and ensure nutritional needs are met for not only the destitute, but children, retirees, and, increasingly, our evaporating workforce.

The ability to forage in the city also brings benefits to those for whom hunger is not a problem, and who have the financial wherewithal to not only eat well, but dine out often. Restaurateurs need regular access to high-quality, low-cost food to remain competitive. It matters little if the restaurant is a tony venue in the heart of downtown, or a mom-and-pop on the commercial strip; profit margins in the restaurant business are tight. The ability to offer diners the highest-quality food while controlling costs to maintain profits can be difficult. Supermarkets offer low-cost food, but generally of lower quality. Farmers' markets have high-quality food, but often at a premium. A system of public produce could provide mutual benefits for restaurateurs and diners, by ensuring a supply of low-cost, high-quality food. And for many restaurants, it is the use of locally grown food that distinguishes them from other eating establishments. Ambiance and culinary talents may not be sufficient in today's competitive restaurant market. It is the quality of the food that is becoming increasingly important, and many patrons now judge quality based on the distance food has traveled from the field to their plate. Being able to forage in the city for quality ingredients satiates the discriminating diner's appetite for fresh, locally grown food while cutting overhead costs for the restaurateur.

It is interesting to ponder a symbiotic relationship between restaurateur, forager, and city government through a program of public produce. At the core of our most vibrant and convivial downtowns and urban neighborhoods are restaurants, cafés, and other eating establishments. Indeed, a popular strategy for economic development in cities throughout the nation is to seed urban places with diverse places of food consumption. Even developers of today's suburban shopping centers are seeking eateries—not department stores—to anchor their developments. Eateries provide entertainment throughout the day and well into the evening, and attract repeat customers. They add to the culture and nightlife of the city, and the best neighborhoods are imbued with generous helpings of eateries. It seems natural for cities to want to guarantee the financial success of eateries, and if locally grown food is becoming a requisite for today's menus, than it certainly behooves the city to ensure such food is within city limits, at low cost.

I remember vividly an excellent *pastificio* in Berkeley's "Gourmet Ghetto" that went to what many would consider extraordinary lengths to secure high-quality, locally sourced produce at a reduced cost. An artfully, hand-drawn sign prominently displayed in the storefront read, "Wanted: Meyer Lemons for Trade or Purchase." As Meyer lemon trees are fairly common in residential gardens throughout the Bay Area, I went inside to inquire a bit more about the offer. The counter person informed me the pastificio uses Meyer lemons in a variety of recipes, such as the specialty pastas and baked goods they produce. Many chefs and gourmands prefer this lemon variety for its distinctive, slightly sweeter flavor. Because of demand, Meyer lemons are difficult to obtain in quantity from local grocery stores and those that do carry them charge a premium.

"If a person comes in off the street with a box of Meyer lemons, you would buy them?" I asked.

Meyer lemons—a favorite of gourmands—command a premium price and are difficult to obtain in quantity. This Berkeley, California, *pastificio* offers to purchase neighborhood Meyer lemons at a modest cost or barter for baked goods, in order to reduce overhead costs.

"Yes," the woman at the counter replied. "If the person wants cash, we pay a dollar per pound. Or we are happy to offer baked goods or other menu items in exchange for the lemons."

I immediately realized there could be mutual benefit with this particular offer, between the restaurant and a forager less fortunate. I was suspicious, however, believing that what the owners may have unconsciously envisioned when they posted the sign was a well-to-do, middle-aged woman waltzing in with a box of Meyer lemons from her backyard. Thinking there may be an opportunity here for those hit hardest financially, I asked the young lady, "What if a disheveled street person came in with a box of lemons? Would you still buy them or offer food in exchange?"

"Not necessarily . . ."

"Why?" I interrupted. "Are the lemons somehow unfit for human consumption simply because they have been handled by a street person?!"

"No, it's not that," she answered. "They can't be any lemons. They have to be *Meyer* lemons."

Obviously this offer is unique, but I held a new appreciation for Meyer lemons as the most prized of citrus fruit. As I left, I couldn't help but think that if only there were Meyer lemons to forage from the urban environment, the impoverished would not have to dig through the trash, or beg for another fast-food burger for a meal. Instead, that person could enjoy delicious, freshly prepared food in the warm and cozy atmosphere of this Italian bakery.

As good as the pasta and baked goods were from this eatery, rents and other overhead costs proved too high, and the pastificio closed its storefront. The sign asking for Meyer lemons, though I had not recognized it at the time, was a public plea for help. I do not know to what extent a regular supply of low-cost Meyer lemons could have altered the ill fate of the pastificio, but I like to think that any opportunity to cut overhead costs could have saved this restaurant, or at least increased its longevity. As it is, the community lost a beloved business.

Not long after the pastificio shut its doors, I came across two healthy Meyer lemon shrubs growing in the public right-of-way between the sidewalk and the street, not more than three blocks from where the pastificio once operated. Remembering their unique cash or barter offer, and thinking of the mutual benefit that could exist between restaurateur and forager, I looked at these lemons in an entirely different light. For some, those yellow fruits are as good as gold.

What I have come to conclude is that within this broader category known as urban agriculture, there are inherent communal benefits of produce grown on public lands that are thus available for public consumption. Public produce that is not only free for all, but may actually yield a monetary return for those willing to harvest (as was the case with the pastificio), creates a symbiotic relationship between restaurateur and forager. In other words, the fruits of one's labor could have reciprocal value. The reciprocity is not only between these two individuals, however. It extends to the dining patrons, who also benefit from high-quality produce at reduced costs, and to the public at large. When a beloved business is able to trim overhead costs and create some financial stability, maintaining a longer, stronger presence in the neighborhood, the entire community benefits.

Some restaurateurs forage not out of financial necessity, but out of principle. Chez Panisse, the internationally acclaimed restaurant and shining star of Berkeley's "Gourmet Ghetto," is certainly in no danger of closing its doors anytime soon. Its popularity and success has sustained for four decades, owing to the culinary talents and slow-food philosophy of Alice Waters and her talented staff. Serving the freshest locally-sourced food is a large reason for the restaurant's success, and has helped grow Berkeley's reputation as the foodie capital of the country.

Staff at Chez Panisse regularly forage for ingredients for their daily menu items. Many of the tarts, pies, and pastries prepared by the restaurant are filled with fruit foraged from various neighborhoods—fruit that is either difficult to source even from farmers' markets, or particularly expensive. According to Stacie Pierce, a pastry chef at Chez Panisse, as much as 30 percent of the fruit used for pastries is foraged. Meyer lemons, blood oranges, huckleberries, kiwi fruit, bitter almonds, black walnuts, persimmons, passion fruits, kumquats, pears, apples, blackberries, mulberries, and raspberries are a few of the foraged fruits that find their way into Stacie's pastries. Some of the items are brought in by locals. Stacie recalls a woman who was hiking in Santa Cruz, and showed up at the restaurant's door one day with a basket of huckleberries. Others arrive with foraged mushrooms from the Berkeley hills, like the prized, golden-colored chanterelles. What has always distinguished Chez Panisse from other restaurants is its flexibility—the ability to change the menu based on not only what is available, but at its freshest and most flavorful. "If you have found something truly amazing," Stacie says, "you don't have to ask if we can use it. We *will* work it into the menu."[13] Though the restaurant readily accepts foraged food from locals, staff prefer to forage for food themselves. These culinary artists have developed the talent to recognize when food is at its peak of flavor. Sometimes, the restaurant has to turn away foraged food brought in from neighbors, simply because it does not meet the staff's high standards for quality. But through rejection, the forager becomes better educated on food and food quality and, over time, develops an appreciation and keen sense of food usually mastered only by talented chefs.

While even restaurateurs and patrons of critically acclaimed restaurants can benefit from food foraged within their urban surrounds, their sustenance is more psychological than physical. Locally sourced produce gives diners satisfaction largely based on environmental principles, beliefs toward improved health, or notions of improved flavor. Some in this country have the luxury of choosing what, when, and how often they eat and where their food comes from. Others do not possess such luxuries. Opportunities to glean and forage for food offer society's less fortunate—those with limited options in life—a choice. While we are starting to see a growing number of middle-class folks who benefit from gleaning and foraging opportunities in the city, the biggest benefactors will always be the most financially challenged.

Back in Berkeley, a man sits idly near a mature redwood tree in Ohlone Park. Richard is homeless, and sleeps every night under the relative protection of the tree's canopy. His behavior and demeanor are not what we typically associate with homeless people. Richard is well-spoken, congenial, and, all things considered, well-dressed. He doesn't appear to suffer from any psychological or substance-abuse problems, and there are never any empty beer cans or liquor bottles near the area where he sleeps. Every morning, after Richard awakes and stretches, he stuffs his bedroll in a backpack, along with some other belongings, and heads downtown in search of food and social contact.

Richard, like many homeless people, is a middle-aged person with limited education, few skills, and no family. Nevertheless, he is articulate, well-mannered, and wise in ways that well-healed college graduates may never be. Recalling Fallen Fruit's mission, I ask Richard if he would find it desirable if fruit trees were planted in the park. "I think it would be," he replies. "There are many days I wish there was just a peach tree around here."

I find it curious that it is not necessarily "things" that people with little in the world crave, but experiences. For Richard, it wasn't just any fruit, but specifically a peach. It has been a long time since Richard tasted a peach, and it became evident that he wasn't just craving food for simple sustenance, but the experience it yielded. A peach's flavor, aroma, and its juiciness provided fond memories for Richard, memories of when he was younger—memories of when he was in a better, more stable situation in life.

We cannot underestimate the value of enriched experiences in our daily lives (recall Solomon's essay *Peaches*). Picking fruit from a tree is more enriching than buying it from the supermarket—both spiritually and financially. For Richard, the ability to pluck a peach off a tree is more than convenient sustenance. It is the opportunity to return to some semblance of self-sufficiency, by not having to rely entirely on handouts from others, while recapturing an innocent joy of youth that has value for Richard. He now worries he may never taste a peach again, because of his particular economic station in life. But Richard's real problem isn't his lack of money. Rather, it is the modern ways that agriculture is produced in this country and made available to people in the city. A person's depleted finances should not prohibit him from eating a fresh peach.

Richard mentioned that the bulk of his time each day is spent searching for food. It was then that I realized, for some, time is just as important as money for those deepest in debt. Searching in trash bins or begging for money for meals is very time consuming, and often yields food that is at best palatable, and at worst, harmful. Many homeless, like Richard, could have more time to seek employment if they did not have to spend so much time in search of food. Having fresh produce readily available and accessible provides not only more healthful sustenance, but freedom and time to pursue other necessities in life.

I talk to Richard more about the idea of planting a variety of food-bearing plants in public spaces throughout the city. He points to the community garden in the park, "Well they have a vegetable garden over there. The only problem is that nobody like me can get in." Richard had an interesting point. The posted sign reads, "Ohlone Community Garden." Along the bottom of the sign are hastily painted letters, "*No Entrar.*" The garden, surrounded by a five-foot-tall

This community garden, located in a public park in Berkeley, is locked behind a chain-link fence with a padlock, and for added security, a combination lock. Most disturbing is the racially biased directive "*No Entrar*" painted across the bottom of the sign.

chain-link fence, has only one access point. This point of entry is secured with a padlock and, for extra protection, a combination lock. Obviously, food here is highly valuable, and some members of the community are doing their best to keep other members of the public out of community plots on public land. What is most disconcerting is the racial bias to the message. One wonders whether there is proof that Hispanics are responsible for stealing the produce, or is it just an assumption?

It is in these instances when one realizes that perhaps community gardens are a misnomer. Though they may be located on public land, the typical community garden only benefits a few individuals—individuals who sometimes go to great lengths to keep the "community" out of their community gardens. Public produce should benefit all by providing for all, where food grown on public land is not locked behind fences, but is freely accessible and available to everyone. One of the greatest shortcomings—and ironies—of traditional community gardens is their personal, privatized nature. The people who typically benefit from community gardens are only those individuals who put down a monetary deposit for a plot, pay for their own plants, fertilizer, and compost, and who take the time to sow, tend, and harvest the gardens. It is understandable why most would choose not to share food with others who do nothing to help purchase, plant, or maintain the garden, and many will argue this is only fair, and how it should be. America is not a socialist country, after all. But America has a growing population of have-nots, and the strength of this country is

The privatization of community garden produce is sometimes blatantly—and at times, ridiculously—obvious. The fact that the public pilfers the veggies underscores the need for opportunities to forage.

directly tied to the health and wealth of *all* our citizens. Surely, individuals should have the freedom to garden, and the right to keep everything they sow and reap for themselves. But we must also recognize that *The Little Red Hen* ethos of community gardening does little to benefit the greater community.

It is important to take a moment to point out other limitations, or at least misconceptions, of traditional community gardens, because they are often regarded as the ultimate safety net for many of our food problems. In his more than thirty-five years of urban agriculture experience and community food service, author Mark Winne admits that our often idealistic claims of self-reliance through community gardening "come precariously close to self-righteous pontificating." Winne explains that "having witnessed the many sincere but ultimately failed attempts to transform dirt, water, and seed into food, I tend to look somewhat askance at those who suggest that more of us, if not all of us, and especially the poor, should 'grow their own.'"[14] Winne's point is that many fail to recognize the effort, knowledge, and resources that are necessary to grow food. There will always be citizens that either lack the skills to grow fresh produce (children, for example), the time (because they work two jobs to make ends meet, attend night school, or are single parents), the strength and dexterity (many elderly and disabled), or the financial resources to purchase seed, soil, and tools (the impoverished). For these individuals and others, there need to be opportunities to forage.

Another shortcoming with traditional community gardens is that they allow municipal government to appease a persistent citizen group without much effort on the government's part. In many communities, the local officials do little more than give permission to a group to plant vegetables on city-owned land—often on vacant lots the city has acquired that nobody wants anyway. Sure, the municipality might pay the water bill, and perhaps offer compost, but these products are not paid for collectively by the taxpayers; they are often paid from the fees that municipalities levy against citizens wishing to garden. Some municipalities do not even want that level of involvement, which often necessitates a third, not-for-profit party, like a Friends of the (insert name here) Community Garden. These groups are responsible for securing funding, managing the supplies, paying the water bill, policing, and other efforts, with little to no assistance from the municipality. In short, many forms of community gardening represent a very hands-off approach to urban agriculture for the municipality. What is necessary, I am arguing, is for municipalities to adopt a more proactive, hands-on policy.

Community gardens are undoubtedly beneficial to cities, and are currently the largest component of urban agriculture today. There is no disputing the good that is intended with community gardens, and land should be set aside for more of them (as long as cities do not try to take that land away at a later date). At the heart of any successful urban agriculture endeavor, now and in the future, are community gardens. These will most likely supply the greatest diversity of produce, but such diversity can require the most labor. We need to recognize, however, that community gardens alone cannot feed an entire community, as their semi-private nature eliminates any possibility for people (other than those tending the plots) to forage. A balance needs to be struck between community gardens on public land (maintained as if they were private), and *true* public produce, meaning food available to all. Public gardens, like those Tom Flaherty cultivates (outside the Parking Office in the City of Davenport, maintained by him and his crew), will have to supplement conventional community gardens. The role of city government that endorses the concept of true public produce is to manage and tend public gardens with available municipal resources and talent, or else hire the skilled and the learned to ensure the health and well-being of the plants, which will ensure the health and well-being of the community.

Some argue that providing access to healthy, low-cost food is not the role of city government. As long as city planners and elected officials strive to create programs to reduce social inequity, and increase the quality of life for their citizens, I contend it is. For the same reasons that city governments provide clean drinking water, protection from crime and catastrophe through the establishment of police and fire departments, sewage treatment, garbage collection, fallen-tree disposal, and pothole free streets, access to healthy, low-cost food helps assure the health and safety of the city's citizens. The surest manner to provide nutritious, affordable food for those citizens is to create opportunities to glean and forage in the city.

Poverty is likely intractable, but hunger does not have to be. Public officials need to recognize hunger's pervasiveness across the country, and fight to eliminate it. Programs and policies need to be crafted and resources set aside to ensure that *all* of life's basic necessities are met: health care, shelter, clothing, as well as food. Gleaning and foraging for fresh produce can directly meet one's needs for food, and may indirectly help them meet the other three life necessities. Eating healthier can obviously reduce the number of illnesses and subsequent doctor visits attributable to poor diet, and if food can be had for little to no cost, enough money may be spared to help purchase clothing, or even make rent. Some individuals, to be certain, are beyond the financial means to acquire additional clothing or shelter, even if there was an extensive system of public produce. But for many single-parent families, elderly, and working poor, a fine line is walked between solvency and ruin, and every penny helps.

CHAPTER FOUR

Maintenance and Aesthetics

Fruit and nut trees are illegal along the streets of most cities. Their crime: producing nutritious food that can fall with a squish into the public domain.

Richard Register, *Ecocity Berkeley*[1]

The biggest objections to planting food-bearing plants in public spaces have always been, and will likely continue to be, maintenance and aesthetics. Public officials are quick to point out that edibles are messy and difficult to maintain, precluding their use in the urban environment. Others contend that vegetable and fruit gardens are a bit unkempt, or even downright ugly, and thus inappropriate in public settings. These concerns are often based largely on misconception and subjectivity. Still, many of these concerns can be addressed with an understanding that maintenance and aesthetics can be balanced by choosing certain plants over others, mixing edibles with ornamentals, utilizing existing maintenance staff and methods, and properly gauging community demand for fresh, local produce.

The pretense that edibles are inherently messier than ornamentals is pervasive, and needs to be addressed. The common perception is that the mess edibles leave is not only cause for aesthetic alarm, but public liability as well. While some varieties of fruit and nut trees provide a basis for these concerns, and only then in certain public settings (and only then if nobody harvests the food before it drops), there is hypocrisy over edibles being inappropriate plants for the urban environment. For any system of public produce to be effective, public officials and those caring and maintaining our landscaped grounds need to take a more critical—and objective—look at the varieties of plants commonly planted in our public spaces.

During my tenure as a landscape architect, I have witnessed a common misperception that the ornamental plants frequently used in our urban surrounds are generally clean, tidy, and maintenance free. Nothing could be further from the truth. Once we look with a keen eye toward the nature in our cities, it becomes evident that the plants commonly found in our public spaces are just as messy—sometimes more so—than many of our familiar food-bearing plants. For example, many would argue that apple trees are too messy for public spaces. Ornamental flowering plums (*Prunus cerasifera*), on the other hand, are widely considered a fantastic addition to any urban landscape, and are frequently planted in cities throughout the country because of their brilliant wintertime floral display. Some cultivars even have purple leaves, making them highly desirable because of their unusual foliage contrast. Many varieties of the flowering plum do produce fruit, though it has almost no edible value (the fruit is only about one inch in diameter with a largish stony pit). While many praise the unique foliage color of some varieties, and the magnificent beauty of the blossoms, the fruit drop can be extremely messy, as the flesh and juice from these little plums are quite effective at staining not only pavement, but the hoods of cars as well.

The ornamental flowering cherry (*Prunus serrulata*), like its cousin the flowering plum, is also prized in the urban landscape. Yet it, too, produces an abundance of small, inedible fruit, which poses both a maintenance burden and liability risk during fruit drop. But its showy blossoms ensure that flowering cherries remain prized plants in the urban landscape. Another example is the strawberry tree (*Arbutus unedo*), which is very popular in California and throughout the drier regions of the West. This large shrub or small tree possesses gorgeous clusters of multicolored fruit. Scores of little orange, red, and yellow drupes dangle from dark-green, leathery leaves, providing reason why the strawberry tree is a favorite with Western landscape architects. Though the fruit can be made into jellies, it is generally considered too bland and mealy to be palatable. These visually striking fruits are soft and squishy, turning light-colored concrete into a darkened and stained eyesore.

Victorian box (*Pittosporum undulatum*) is a beloved urban street tree, widely used in cities with temperate climates. The thousands of orange berries that each tree produces, though visually stunning, provide a sticky mess, frustrating both pedestrians and car owners unlucky enough to park their cars under the tree during fruit drop. *Cotoneaster* and firethorn (*Pyracantha*) are favorites among landscape architects because their profusion of brightly colored red and orange berries shows off a fantastic display. But these berries provide only eye candy, as they are unpalatable to humans; and yes, they, too, make quite a mess. Yellow pine pollen coats everything within a breeze's reach, and acacias not only aggravate allergies, but also require sidewalk cleanup during flower drop. *Pittosporum*, bottlebrush (*Callistemon*), and *Jacaranda* drop flowers with sticky nectar. Leaf litter is a problem with Chinese elm, redwoods, pines, and cedars. Sweet gum (*Liquidambar styraciflua*) and Red Horsechestnut (*Aesculus* x *carnea*)—popular street trees in the South and Mid-Atlantic that are prized respectively for their brilliant fall color and spring blossoms—drop dozens of hard, one-inch-diameter seed capsules to the sidewalk and street. These capsules present a safety hazard and potential liability, as people could slip or roll their ankles on the round pods.

The selection of trees and shrubs commonly used in urban surroundings begs the question: In the name of aesthetics, can we—or more to the point, should we—justify using these messy species that produce an abundance of leaf-litter, drip with sticky nectar, or drop unpalatable fruit by the bunches? It seems their incredible aesthetic value makes it easy for us to forget the associated mess that accompanies these plants. Persimmon, fig, Asian pear, lemon, banana, orange, pomegranate, almond, and scores of other food-bearing plants,

Sweet gums (*Liquidambar styraciflua*) are prized street trees across much of the United States. The scores of hard, round seed pods dropped from each tree arguably provide as much risk to pedestrians as fallen fruit and nuts.

however, possess equally showy aesthetic qualities, in addition to their delicious and nutritious fruit. Passion vine, for instance, is an excellent alternative to trumpet vine, offering a more exotic-looking flower and wonderfully aromatic fruit.[2] Grapevines that are trained along white picket fences, or ramble over pergolas, trellises, and arbors, also provide both beauty and sustenance. They make wonderful substitutes for wisteria—a plant often used for its fragrance and visual display of long, drooping clusters of flowers. Instead of clusters of flowers, think clusters of berries; one titillates our sense of smell, the other our sense of taste. Wild strawberry provides a vigorous, low-maintenance ground cover. The berries that do not get harvested usually disappear under the mat of green leaves, inconspicuously decomposing, adding fertile compost to the topsoil. Kale, cabbage, sorrel, Swiss chard, and a host of other leafy greens provide tidy, no-mess additions to any perennial bed. Ornamental grasses, such as purple fountain grass and switchgrass, among others, have become quite popular in landscapes, providing a lacy, feathery accent that hints at wildness. These grasses are often praised by environmentalists because of their drought tolerance. Fennel can provide a similarly lacy, wild look, and is every bit drought tolerant as ornamental grasses. Rosemary, lavender, and thyme are about as maintenance-free as plants get, providing drought-tolerant alternatives that are handsome, fragrant, and edible.

There are also degrees of "messiness." What is worse, a conifer that dribbles sticky sap and drops needles throughout the year, or a deciduous tree that releases an abundance of leaves all at once? Is an apple tree that drops fruit once a year more of a maintenance headache than a silver maple that heaves and breaks sidewalks at maturity? These questions are difficult to answer definitively. The real issue, rather, is that almost all trees and shrubs are messy, and fallen fruit, branches, wet leaves, sticky flowers—regardless if the plant is ornamental or edible—pose some aesthetic affront and potential liability.

It is therefore difficult to indiscriminately dismiss food-bearing plants in public landscapes. We should recognize that plants provide a greater good than simple aesthetics. For municipalities that accept a philosophy that food security

Passion vine provides an aesthetically pleasing—and exotically edible—screen to an otherwise unfriendly chain-link fence. The particular variety shown here, banana passion fruit (*Passiflora mollisima*), can be grown in temperate climates, making it more useful in many parts of the country than its subtropical cousin, purple passion fruit (*Passiflora edulis*).

is one of those greater goods, attention should turn to strategies that can feed hungry citizens without placing an undue burden on maintenance staff.

Richard Register, the notable environmental planner and urban theorist who has spent decades investigating and promoting the food production potential in cities, understands well the reticence people have toward edible plants in public places. In his book, *Ecocity Berkeley: Building Cities for a Healthy Future*, Register examines the role that streets, for example, could play in providing fresh produce—namely, fruit and nut trees planted in the public right-of-way between the sidewalk and curb. He explains why public officials across the country are loathe to plant these particular edibles in the urban environment, and offers suggestions to reduce maintenance burdens associated with food drop:

> Fruit and nut trees are illegal along the streets of most cities. This is because some owners fail to harvest or clean up under their fruit trees, thus creating an aesthetic offense in other people's eyes and a liability if someone were to step on a fruit and slip. Both objections are legitimate but could be reduced in two ways. First, establish a legal procedure for taking responsibility for the trees: either the city hires a roving orchard farmer (city employees presently trim ornamental street trees in any case), or the landowner who wants the trees accepts responsibility for upkeep, and for liability and penalty. In this same spirit, the food-tree lover could strategically plant trees with soft fruit (plums, peaches, and some pears) where they do not overhang sidewalks, while reserving nuts and harder fruit (apples and lemons) for the more public locations. Second—and this is a matter of degree—people should take responsibility for themselves. To slip and fall on a sidewalk because of a fruit should not yield a gigantic settlement for the unalert person—a small settlement based on shared responsibility would make sense. Perhaps people who can't trust their own awareness should insure themselves; laws certainly should encourage transfer of a good deal of responsibility back to the individual.[3]

Public officials' concern over food drop and liability is real, and Register's strategies offer a couple of methods to help protect the municipality. There is no doubt that municipalities will have to devise innovative methods to limit the amount of food that drops to the ground, and is thus wasted. But such a challenge should not discourage cities from exploring the greater benefits of helping to establish food security for its citizens. In some cases, the strategies present a win-win for the community and the public officials.

One simple solution could prove to be the most effective. A sign, artfully written and conspicuously placed, encouraging people to harvest food may be all that is needed to minimize waste. Many are reluctant to harvest edibles because they are often perceived as belonging to somebody else. We are not used to seeing food growing right before us in our urban spaces, and this change in context—from one where we give somebody money for food to one where it is offered free of charge—will take some getting used to. Until then, a simple message of encouragement can help ease people's reluctance to harvest.

Of course, some situations may call for more creative strategies to minimize edible waste. The University of California at Davis has devised an ingenious—and lucrative—method of dealing with fruit drop, providing a lesson that has value to any municipality. Olive trees—close to 2,000 of them—line walkways, bike paths, and other public spaces on the Davis campus, creating not only a maintenance headache for the grounds crew, but a real liability for the school as well. Navigating through the squished olive fruits and stony pits requires considerable caution. It is manageable for pedestrians, but in this bicycling community, the fruit drop can prove treacherous to those on two wheels. In 2004, sixty thousand dollars was spent in legal fees for bicycle accidents related to olive drop. This was in addition to the annual cost of another sixty thousand dollars just to clean up and dispose of the olive fruit. It was a never-ending nightmare for Sal Genito, director of the university's Buildings and Grounds Division. One day, after he was called to the site of a particularly bad bicycle accident, he had an epiphany. Genito was surveying the scene of the accident alongside Russell Boulevard, an area where hundreds of olive trees line the space between the bike path and the street. Smashed olives were everywhere, creating an extremely slick surface that was made even more treacherous with light rain. The smell was inescapable: "Olive oil," recalled Genito. He then bought a small press, picked some fresh olives, and churned out a fragrant and delicious green-hued liquid—and UC Davis Olive Oil was branded.

In the fall of 2004, the first olive-oil vintage of UC Davis, eighty gallons of artisan extra-virgin olive oil was pressed and bottled. It was an immediate hit with consumers, and production has increased substantially since. The 2006 vintage yielded almost 450 gallons, and sold out in just four months. The 2007 vintage produced close to 800 gallons of oil (which, by the way, comes in three distinct blends, depending on the varieties of olives used). At twelve to fifteen dollars per 250-milliliter bottle, the financial returns are staggering. The 2007 vintage generated close to eighty thousand dollars in profit, which helped support the UC Davis Olive Center, a new education and research facility devoted to the production of olives and olive oil. A simple yet ingenious idea that stemmed from a maintenance headache, UC Davis's olive-oil program not only generates enough revenue to cover maintenance and liability costs of fruit drop on public space, but now fosters widespread understanding and appreciation of this gustatory delight.[4]

A growing number of consumers today seek local food, or at least demand to know the origin of the food they eat. People are getting wary of food produced in far-off regions, and, ironically, those who can afford to will typically pay more for food grown locally. It is time to recognize that locally produced food has immense value. UC Davis's olive oil and the City of Chicago's rooftop honey are just two examples of innovative strategies that turn maintenance crews into moneymakers by exploiting the demand for locally produced food. Similar strategies merit consideration, not only for fresh produce, but for the various dried fruits and vegetables, preserves, jams, jellies, nut and seed butters, oils, relishes, and other value-added foodstuffs that could be produced by the municipality and sold back to the community. In an effort to manage food drop and to offset maintenance costs, there may be many opportunities to provide food directly to local citizens and receive some monetary compensation for the efforts.

The surest manner to reduce risk and burden from food drop is to simply guarantee little to no waste. This requires managing the food supply to match consumer demand. In other words, maintenance can be minimized when we effectively estimate the carrying capacity of each agriculture activity: How much food *should* public space produce, rather than *could* it produce. Municipalities interested in experimenting with edible landscapes need to consider the number of people occupying, or at least passing by, each and every public space, and whether those people are likely to consume the quantity and type of food that could be offered. Lining a suburban street with persimmon trees spaced forty feet on center, for example, will likely yield an overabundance of fruit for the relative paucity of residents in the subdivision, especially if those residents have never developed a fondness for persimmons. A single peach tree in a busy, inner-city park, on the other hand, may not yield enough fruit for hungry individuals. Matching expected crop yields to numbers of people likely to harvest the produce is paramount in reducing management headaches of urban agriculture.

The importance of proper carrying capacity with regard to edibles in public space became evident while I was living in Berkeley a few years back. What caught me by surprise was the general lack of fruit on the ground from the many neighborhood fruit trees. Many of Berkeley's neighborhoods are densely populated, and even along the "quieter" streets, an abundance of people pass by on foot or bicycle regularly. These neighborhood streets superbly illustrate the feeding potential of food-bearing plants in the public realm, while also addressing the maintenance concerns of city officials.

On Grant Street, a few blocks northwest of downtown, a fifteen-foot-tall navel orange tree thrives in the narrow planting bed wedged between the sidewalk and curb. It was the middle of February, and dozens of good-sized fruit were ripe and ready to pick. Interestingly, only a couple of oranges lay in the gutter, and most of the fruit that would ordinarily be within an arm's reach had already been harvested. It seemed that this tree had been providing winter treats for folks. Similarly, on Channing Street, a few blocks south of that navel orange tree, a fig tree also thrives in the confines of the public right-of-way between the street and sidewalk. When the figs are ripe and ready for harvesting, few fruits are ever found on the ground. Figs are prized fruit, similar to oranges, and I suspect this is partially why there is little waste. Since these two neighborhood trees are in the public right-of-way, the fruit they bear does not belong solely to the people who planted them, even though they care for them. The fruit belongs to everyone, gifts from private individuals to the public at large.

Two orange trees in two cities in California. The remarkable abundance of ripe fruit on the trees, and the almost complete absence of wasted fruit on the sidewalks below, prove that public produce is prized in some communities.

One day, as I was admiring the fig fruit, the resident who cared for the tree emerged from his house. "Help yourself," he kindly offered. "They didn't do so well this year, but they're still okay, and we can't eat them all anyway."

I asked him when he planted the tree. "Five years ago," he responded.

"Have any of the neighbors or city people ever complained about fruit drop, or mess during that time?" I inquired.

"No, not at all. There's really not that much that falls to the ground. We pick some, but I think other people occasionally pick the fruit as well. Actually, the leaves make more of a mess during the winter when they all fall off, but nobody complains about that either."

Three blocks west of downtown, between the orange tree on Grant Street and the fig tree on Channing Street, a few homeless individuals sleep in Ohlone Park. As with many urban open spaces, this neighborhood park is an attractive place for the homeless. But this park also attracts dog owners, teenagers, parents with young children, and other diverse members of society, and thus provides an interesting case study.

Amid all the people, dogs, benches, drinking fountains, and other park paraphernalia, an apple tree grows near an exercise station along the park's par course. In November, when the apples are ripe, few fruits can be found littering the grass below. The canopy of the tree still houses scores of hanging fruit, but most are outside of a normal person's reach. But soon, even the fruit high up in the tree disappear. Like the orange and fig trees observed in the neighborhood, this tree seems to be feeding folks as well.

The fact that fruit is eaten without waste littering the sidewalks and streets in Berkeley demonstrates proper carrying capacity for each produce item in its particular context, and exemplifies the many factors that must be weighed in determining carrying capacity. Of course, there exists a strong food culture in Berkeley that prizes locally grown, organic produce. Understanding the degree to which locals demand access to fresh, affordable produce is essential to estimating carrying capacity. The relatively high population density in the Berkeley neighborhoods and the number of people passing by the fruit trees also has much to do with near zero waste. Visibility is also important in managing food litter: *Where* food is planted within a particular space is just as important as how much food is produced. If edibles are planted in a back corner of a seldom-used park, expect lots of waste. They should be prominently displayed in the landscape, reminding people of their food choices, and inviting them to harvest. All of these examples in Berkeley were in plain sight. It was obvious to anyone passing by that the fruit trees were planted in public space, accessible from a sidewalk and thus available to all, regardless of who tended them.

Not all food drop is objectionable. In fact, without some food drop, many passersby may never notice the bounty of food overhead. I find it interesting that even today, as an experienced landscape architect and urban designer, I often fail to notice a particular species of tree I may be passing unless I spot a clue on the ground. We often just perceive trees as leafy green masses on sticks, pleasant for sure, but somewhat homogeneous. It isn't until we see a walnut on the ground, or an apple, that we stop and look around for the specific source of food.

Managing food drop is arguably the most worrisome issue with regard to edibles, but it is an issue that generally needs attention only once each year. Keeping edible landscapes healthy and thriving year round can be more perplexing to maintenance crews. It involves continual irrigation, mulching, weeding, pruning, and pest management. Public officials will perennially argue there is simply no budget to hire gardeners to maintain a network of public produce. Those arguments seem particularly valid during economically depressed times. Nevertheless, if the goal is not to sell value-added food items back to the community to help offset maintenance costs (à la UC Davis and the City of Chicago), but to provide fresh produce free for the taking, planting more public produce does not typically necessitate hiring additional maintenance staff dedicated to caring for the edibles. Instead, municipalities should tap the skills of their existing workers within a broad range of government departments. Most cities, for example, have a department of forestry (or a forestry division within a parks or public works department) that is already tasked with the maintenance and management of the urban forest and understory shrubs. These folks are often certified arborists and understand the intimate needs of all sorts of woody plants, ornamental as well as edible. Parks and recreation staff provide year round maintenance to park and plaza landscaping and could thus effectively manage edibles while they keep ornamentals thriving. In Davenport, Iowa, Tom Flaherty alone maintains his postage-stamp-sized garden outside his Parking Office window. For the half-acre garden he is planning for a seldom-used parking lot, he will use his existing maintenance staff for the upkeep, as they already maintain the landscaped grounds around the city-owned parking ramps and surface lots. The beauty of public space is that there are already dedicated municipal staff assigned to the upkeep of the landscape, meaning the care of food-bearing plants can be done by the same employees that tend to the rest of the plants. The skill set is no more complex than golf course maintenance. If municipalities can justify the immaculate care of turf and associated landscaping for a recreational sport, perhaps they can find cause to dedicate equally skilled staff to a life necessity.

In the case of municipal and civic buildings, the grounds may not be maintained by city employees, but by an outside contractor. Either way, landscape contractors and city staff are already well-versed in a variety of maintenance methods. Certainly, there are some maintenance contractors that know little beyond mowing lawns, blowing leaves, pruning trees into perfect lollipops, and trimming shrubs into sumptuous gumballs. But many others also know how to plant and stake trees, transplant shrubs, mulch planting beds, weed, deadhead, prune, and fertilize ornamental landscapes. These skills are exactly what are necessary to grow edibles. In other words, the grounds are already being maintained with skilled labor, so it is not a matter of hiring new staff, or even retraining existing staff. It is more a matter of redefining how workers are currently maintaining the spaces.

The one network of public space—and it is a big one—that typically does not have ongoing municipal staff or outside contractors dedicated to landscape maintenance is streets. Street trees and other plantings along residential streets, for example, do not generally receive any municipally sponsored maintenance or irrigation. Upkeep is often left entirely to homeowners. Property owners may have been the ones to plant a tree and other landscaping in the first place, so they have an inherent interest in the maintenance of the plants. Even if the city decides to plant fruit trees, maintenance policies could still be the responsibility of the homeowner, as Richard Register suggests. The requirement to maintain a clean sidewalk is no different than, say, snow removal. In communities throughout the Midwest and Northeast, it is the homeowner's responsibility to remove snow from city sidewalks. Homeowners that fail to comply will have the walk shoveled by the city, at the owner's expense. The same policy could be implemented during fruit drop, a shorter time of the year than winter. But as evidenced along Berkeley's neighborhood streets, such a policy, at least on residential streets, may not be warranted. Most homeowners, and even a few renters, want to keep the grounds of their living quarters neat and tidy. Understanding this desire to maintain the grounds in front of one's abode, the city could encourage residents to plant food-bearing trees and shrubs—or at least not penalize them for doing so.

In commercial or mixed-used areas of the city, the city government or Chamber of Commerce may want to work with individual business owners, or existing Business Improvement Districts (or help establish one), to maintain edibles.[5] It is in the best interest of the business proprietor to keep the sidewalk outside his or her storefront attractive and inviting, which could be accomplished with edible displays in large planters. An interesting program along these lines is exactly what is happening in Des Moines, Iowa. To help enhance the streetscape and businesses downtown, the Parks and Recreation Department created a program much like one would find within any Business Improvement District. The premise of the program is quite simple: Merchants supply planters outside their storefronts, and city staff supply the soil and plants. Parks employees even maintain the plants, though many merchants often prefer to do so themselves. It should be noted that in the case of Des Moines' program, the plants are ornamentals. But such a program could easily be instigated for ornamental edibles, such as herbs, kale, Swiss chard and other attractive leafy greens, or more compact varieties of produce, like trailing strawberries, peppers, or the smaller Japanese eggplant varieties. Such displays not only provide convenient sustenance, but can attract attention from the passerby. And what merchant doesn't want to attract a bit more attention to his or her storefront?

Commercial areas may provide another option for maintenance of edibles—one for municipalities intrigued with the idea of public-space agriculture, but that want absolutely no part in maintaining or harvesting the food. Such a strategy would be to place conditions upon commercial developments, like shopping centers and office parks. The semipublic spaces of these developments—namely, the landscaped grounds around the associated plazas and parking lots—are suitable locations for public produce. Though the land is pri-

Giant herb pots in Austin, Texas, add an engaging element to the streetscape and appear to require less maintenance than the street trees that once existed.

vately owned, the public is freely allowed—and even encouraged—to access the property. Placing conditions on commercial development is commonplace, and something municipal planners and elected officials routinely do. The benefit of edibles grown on commercial development is that the grounds are already maintained by professional landscape contractors. A simple condition of development that could prove a win-win for the municipality wanting to promote a system of public produce, but unable to maintain one, would be to require that 10 percent of all landscape grounds be set aside for edibles with "permission to pass" granted to the public. The condition could further state that the edibles must remain healthy without the use of chemical fertilizers, herbicides, and pesticides, ensuring free, fresh, and organic produce.

Another compromise for municipalities wishing to support public agriculture efforts, but not wanting to maintain them, is to ready public land for others to take over food production. This can easily be accommodated in the Capital

Improvement Program of cities. Given the potential scale of some urban gardens, it may be best if the municipality attempts to restore the land to a fertile, nutrient-rich medium, especially if the land has been developed or paved—as with a parking lot, for example. The amount of compost, the machinery, and the labor required to turn a sterile patch of dirt into a productive medium for growing food is likely too daunting for the small group of individuals that would be growing the food. After the ground is readied, it can then be turned over to the entrepreneurial farmer, neighborhood group, or not-for-profit organization for production and management of food. The one-time labor and capital expense shows commitment by the city for urban agriculture opportunities without placing ongoing maintenance demands on city staff.

Some municipalities have taken a very active approach in the management and maintenance of food-producing efforts by creating specific divisions within existing departments dedicated to urban agriculture. The City of Des Moines and the City of Portland, Oregon, each have urban-agriculture programs housed within their respective Parks and Recreation Departments. Staff in Des Moines distribute plants, coordinate plantings, provide onsite consultation and training, administer grants, and promote programs to further sustainability and food security within their city. The City of Portland staff also help with onsite agricultural efforts, as well as maintain demonstration sites for orchards and small fruit production, and manage garden plots where the harvest is donated to needy people in their city. The City of Seattle houses its famed "P-Patch"[6] community gardening program within its Department of Neighborhoods. City employees help procure and manage gardening sites, and administer programs specifically aimed at providing immigrants, youth, low-income families, and food banks with fresh, locally-grown, organic produce. From New York to Nashville, municipalities are employing skilled and knowledgeable people to help manage and maintain community gardens and other food-production activities on city-owned land.

Instead of creating a division within an existing department, an effective strategy is to house food-production operations entirely within a new department. It has become apparent to many that in the near future, municipalities will need to address food as an important component of urban infrastructure, much like housing, transportation, and education. As such, each municipality would do well to establish a Department of Food that would "embrace urban farming as an appropriate mechanism for accomplishing its mandate."[7] Another option is the nascent and increasingly popular Department of Sustainability. Employees in these sustainability departments are generally concerned with the host of strategies that can reduce the city's carbon footprint, usually through energy reduction by managing electricity and fuel consumption. If a city is truly interested in "going green," as many are, food has to be considered an integral part of sustainability. Sustainability is more than a fleet of hybrid cars, programmable thermostats and light switches, and switching from incandescent light bulbs to compact fluorescents. It is more than reducing the consumption of energy, and it is more than climate change and environmentalism. Sustainability is also about economics and social equity. Environment, economy, and equity are the three legs of sustainability, and food in public spaces provides a footing for all three.

Water is integral to maintenance and sustainability. Regardless of who is responsible for the continual care and upkeep of our public landscapes, the use and cost of water for growing food can be a concern. But, growing edibles does not necessarily require the installation of an irrigation system. Dry farming—the production of crops in arid conditions without the use of irrigation—can be a successful alternative, and is gaining attention across the country, even with commercial growers. This practice is immensely popular with viticulturists, among others. Irrigation increases crop yield, which is not desirable when growing some fruit species, like wine grapes. It is not necessarily the quantity of grapes that the viticulturist desires, but the quality. Irrigating can lead to an overabundance of berries on more numerous clusters. But the viticulturist wants smaller yields, as there will be a greater intensity of flavor and sugar in each cluster, and within each berry. One method to ensure reduced yields is to thin the crops, which is very labor intensive. Another measure is to simply not irrigate (or irrigate less). This strategy works with most fruits and vegetables, and is an excellent way to spot manage carrying capacity. Even a single tree may provide too much fruit for passersby. Reduced yields could better match carrying capacity while reducing maintenance costs and efforts.

If more abundant yields are desired, then irrigation may be necessary. Again, the installation of an irrigation system solely for the production of food is not usually required. In many parts of the country that do not receive adequate rainfall during the growing season (California, Texas, the Southwest, and even in some places in the Midwest, for example) the landscaping in public spaces is already irrigated. In these situations, mixing edibles among the ornamentals ensures that the produce receives a clean and ample supply of water. If the desired food-bearing plants are especially thirsty, they should be planted adjacent to lawns. Turf areas are usually heavily irrigated, and there is generally sufficient overspray to water the nearest plants.[8]

The types of food-bearing plants also have a dramatic effect on maintenance and irrigation needs. Some plants are naturally labor intensive, while others thrive with little attention. Cities would do well to employ principles of permaculture (a portmanteau of *perma*nent agri*culture*); that is, "the conscious design and maintenance of agriculturally productive ecosystems that have the diversity, stability, and resilience of natural ecosystems."[9] One of the easiest ways to ensure resiliency and stability (and thus reduce maintenance demands) is to select plants that are native or well-suited to the particular geography of the city. Native plants often thrive without supplemental water, can effectively ward off indigenous pests (while attracting beneficial critters), and are generally less troublesome than crops foreign to an area. When plants are selected based on their natural suitability to a given locale, the natural order of things shoulders much of the maintenance, creating a productive and sustainable landscape.

There is also a great disparity in maintenance demands between woody perennials and herbaceous annuals. Teva Dawson, community garden coordinator for the City of Des Moines and an active proponent of the municipality's goal to ensure food security in the city, has found better success with food-producing woody perennials. As Dawson notes, trees and shrubs only need care and supplemental water for the first year to get established. After that, maintenance is minimal, thus providing a logical choice for municipalities with strained maintenance resources.[10] Most vegetables, with the exceptions of rhubarb, asparagus, and a handful of other, more exotic varieties, are annuals that require considerable care during the growing season, and will not live to see the next. By contrast, fruit and nuts are produced in abundance on a single tree or shrub, without the need for weeding, fertilizing, or supplemental water. Year after year, trees and shrubs produce without transplantation, replacement, or soil reconditioning. Some years may be less prolific than others; nevertheless, a single fruit or nut tree can supply food for many individuals.

Pest management is a specialized maintenance task, and is of critical concern with edibles. Effectively managing pests in the landscape, even if chemical pesticides are used, can be a time-consuming endeavor for maintenance crews. So, why not let plants shoulder some of that burden? With a bit of plant knowledge, it is easier to manage troublesome bugs and critters without having to place additional demands on maintenance staff. Mixing edibles with ornamentals is a great way to garden organically by naturally managing pests. Picking plants that attract beneficial critters, while repelling malevolent ones, is known as companion planting. It is one effective strategy within the environmentally sustainable process known as Integrated Pest Management. In the city, where there is typically a strong desire for aesthetically pleasing and artfully composed landscapes, companion planting is a no-brainer. One showy plant commonly used in the urban landscape is the marigold. Marigolds are quite effective at deterring all sorts of critters. Their vibrant blossoms are quite attractive to humans, but their unpleasant scent deters aphids, squirrels, thrips, squash bugs, and other pests. Marigolds also release a toxin in the soil that kills nematodes, but is not harmful to humans. Artemesia, or wormwood, is quite effective in deterring many animals and foliage-devouring slugs. Some plants attract beneficial predator insects that, in turn, devour malevolent ones. Alyssum and yarrow attract parasitoid wasps and hoverflies, which prey on spider mites, green flies, and small caterpillars. The plants need not be purely ornamental to deter pests. Plants with pungent scents and spicy flavors are quite effective at repelling unwanted critters, including rodents. Rosemary, onion, peppers, peppermint, garlic, thyme, chives, basil, cilantro, and other piquant herbs and vegetables keep scores of plant-devouring critters in check. Fennel repels fleas, sage repels slugs, and lavender repels mice and moths. It is a miraculous irony that the scents and flavors that are so compelling to people are so repulsive to pests; nature's insurance that we humans can eat, and eat well.

Mixing edibles with ornamentals is not only a marvelous way to reduce maintenance demands but can also ameliorate concerns over aesthetics. Many contend that vegetable gardens and fruit orchards provide an inappropriate look to our public places. We have become a culture where we expect our urban landscapes to be well-groomed. Edible landscapes can be a bit unkempt and, because of that appearance, raise aesthetic objection by some. In the urban environment, it is arguably aesthetics alone that drive the plant palette in landscape design. Talented landscape architects and garden designers, however, weigh a litany of criteria when crafting a plant palette. Some criteria are aesthetics for sure, such as seasonal color, texture, and overall size of each plant for its particular site. But other criteria include environmental concerns, such as drought tolerance or erosion resistance; public safety and comfort (e.g., shrubs that block views from the street, trees that cast lots of shade, or plants with lots of prickly thorns); symbolism and cultural meaning; and, of course, maintenance. All of these concerns, and others, are balanced to enhance public spaces and add value for its users.

There is no doubt that food-producing plants can be messy and require maintenance. Dismissing edibles because they are not as pretty or require much more maintenance than ornamental plants, however, is often subjective and myopic. Sure, intensive row-crop agriculture in our public parks is probably a maintenance headache, and may not be desirable within some urban settings. A challenge that befits the astute public-space designer is incorporating edibles into a successful park design, for example. Just as one would not plant large, thorny shrubs like firethorn and cactus next to a playground for toddlers, so too should a measure of caution be exercised when planting edibles. In other words, sound design principles are not thrown out the window simply because the plant palette uses fruit-bearing trees instead of sterile cultivars. As in any landscape design, the architect needs to take into account how many people will use or pass by the space; what types of activities will take place in the space; the microclimate, solar access, and water availability of the space; and a host of other variables. When planting our backyard garden, we often do not consider these variables. We know we want pole beans, potatoes, radishes, maybe some carrots, and Oh! We have to have *tomatoes* (what garden is complete without *tomatoes?*), maybe some lettuces, and, what the heck, let's try some watermelon. There is little thought given to the garden's aesthetic composition, and there is no need to consider who else might be using the space, since it is in our backyard. This type of edible landscape may be inappropriate in many public settings. A better, more balanced approach is to mix edibles with ornamentals, something we don't typically see in backyard gardens. This means roses with tomatoes, rosemary and citrus mixed with fortnight lily, fennel mixed with purple fountain grass, persimmon and cherry trees interspersed with dogwoods, for example— all within a park or plaza setting that attracts users with beauty and offers opportunities for social and physical sustenance.

A fantastic example of an edible landscape imbued with artistry and beauty was recently completely in Jamaica, Queens, in New York City. A beleaguered community garden, once up for auction in 1999, was spared by a land trust and transformed into a public space brimming with social activity and neighborhood pride. The New York Restoration Project (NYRP), a group founded by entertainer Bette Midler that works to clean and restore various park spaces and community gardens in the city, held focus groups to solicit the gardeners' desires for this particular space. The executive director of NYRP recalled that "none of them liked the way the gardens looked. In some cases, all they wanted was something simple, like a more attractive fence, but in others, they wanted a new design that would make the space feel more open and welcoming." Walter Hood, a gifted landscape architect recognized for his design of bold, sculptural urban spaces, was chosen for the renovation. What Hood created was something few community gardeners had ever seen. Linden trees with an understory of carpet roses announce the garden's entrance, while an arbor covered with trumpet vines runs along the garden's length. Inside the garden, raised vegetable beds are laid in parallel lines, a nod to the rail line that runs alongside the northeast edge of the site. French-styled parterres create formal spaces where boxwood surrounds heirloom vegetables, pumpkins, corn, and other edibles. And perhaps the most striking elements of the garden are the half dozen, ten-foot-tall rainwater collectors that resemble giant blue martini glasses. These colorful collectors funnel three thousand gallons of rainwater to two underground cisterns, providing not only a convenient and ecologically friendly water source for the gardeners, but a sculptural element that attracts attention from all passersby. As Hood notes, "I was trying to find something that might capture the imagination." It seems he has succeeded. As one resident exclaimed, "To me, it's the most beautiful site. All I want is to just sit and absorb it."[11]

We will have to begin to change our perception of what edible landscapes in urban spaces currently are, and recognize what they could be: places to socialize, to decompress, to garden and to forage. And they should also be beautiful. But the idea that food-bearing plants can be both nutritious and beauteous will take time. This is not unlike the shift we have seen from water-intensive landscaping to xeriscaping in many parts of the country. Large clumping grasses and grass-like plants were initially considered too "weedy" for many landscapes. Boulders and gravel often created bleak gardens, and succulents and cacti were inappropriately used in many settings. But with time, we have come to embrace water-sipping landscapes composed by talented designers as not only appropriate in many urban settings, but also beautiful. I believe such will be the case with edibles in the very near future.

What if, after dedicating staff time to ensure the health and beauty of edible landscapes and managing the quantity and quality of food produced, the locals and passersby do not eat the produce, wasting both food and effort? One method to minimize the fruit drop and vegetable spoilage in such a situation would be to enlist the help of volunteers to harvest the largest parcels of food. Cities might be surprised to find just how many people will show up—and how far they will travel—for an opportunity to harvest fresh produce (recall the gleaning event at Miller Farms in Colorado). Volunteers readily maintain and harvest the City of Portland's community gardens that are dedicated to providing food for the needy. Another example of successful volunteer efforts in that city is demonstrated by a grassroots organization known as the Portland Fruit Tree Project. This all-volunteer group works with various neighborhoods to organize "harvesting parties" to ensure fruit doesn't go to waste. In 2007, the group organized eight harvesting parties that were attended by 132 volunteers. These individuals gave of their time to harvest fruit from more than fifty trees in the city, yielding 3,400 pounds of produce that otherwise would have fallen to the ground.[12]

Expanding on the concept of harvesting parties, there may be an opportunity to turn maintenance into cause for celebration. U-Pick farms and ranches, where people drive scores of miles for the opportunity to harvest fresh produce, point to the yearning urbanites have for a bit of agrarianism. Why not organize municipal U-Pick operations for urban orchards and other larger-scaled plots in the city? Like U-Picks, the municipality can charge a per-pound fee for the produce items as a way to partially fund the maintenance program. And, an event that pairs harvesting with other food-festival activities, like cooking demonstrations and recipe contests, can ensure a celebratory spirit and help generate a larger turnout. It can be an opportunity for the municipality to not only guarantee that an over-abundant crop doesn't go to waste, but to build revenue and community pride as well.

Another strategy for eliminating food waste is to donate the harvest. The great thing about food is that it has value to somebody, and some organizations, like food banks, ensure that food gets to those in need. Food banks are not-for-profit organizations that collect food items and glean fresh produce to distribute to food pantries, soup kitchens, and other charitable organizations bent on eliminating hunger. In fact, many food banks will harvest the produce themselves, relieving the burden from the landowner. Other volunteer organizations will grow or harvest food for donation to food banks, relieving municipal staff of these duties. In the case of the Portland Fruit Tree Project, for example, half of the produce harvested goes home with the volunteers, and the other half is donated to food banks. Seattle's P-Patch program produces over ten tons of organically grown food annually for donation to local food banks. While I am principally arguing for a system of public produce that can feed the consumer directly, without the need for distribution "middlemen," growing food with the sole intent of supplying local food banks is an entirely laudable endeavor. In an ideal world (one with a successful system of public produce!) there would be no need for food banks. Yet, with the food security issues this nation is already facing, it is probable that food banks will need to exist for the foreseeable future. Many are seeing a decline in donations, however, as food companies seek to eliminate waste, which reduces company profits. A community agriculture endeavor, like a program of public produce, can bolster a food bank's supply, which guarantees distribution of healthy food to those in the community that need it most.

To the chagrin of architects, planners, and landscape and urban designers, maintenance often drives the design of our buildings and the spaces between them. Rather than this skewed approach, the designs of our human settlements first need to consider the needs of inhabitants, and then bring into play programs and strategies to help assure those needs are met and maintained. As has become obvious, I am focusing on myriad successful strategies to help quell the fears of municipal officials that have long discouraged—or outright forbade—the use of edibles in our urban landscapes solely because of maintenance and aesthetic concerns. I am optimistic that the programs and policies adopted by the few organizations highlighted in this book will help ease those apprehensions. I do not mean to diminish the importance of maintenance and aesthetics in our urban landscapes and public spaces. Beauty inspires us, and proper maintenance plays a significant role in the attractiveness of a space. But these considerations should be balanced with the greater good that providing the community with food choices offers. What typically adorn our urban landscapes, unfortunately, are trees and shrubs that are high in aesthetic value, low in food value, and yet similar with regard to maintenance requirements as comparable edibles. What should be weighed is the added value certain plants provide to users of public space vis-à-vis the perceived added burdens of maintenance. When properly selected, edibles as landscape plants have the ability to achieve all the public-safety, comfort, aesthetic, drought-tolerant, and general maintenance goals required of plants in public space, with one notable exception: They help establish food security.

CHAPTER FIVE

Food Literacy

No matter how you think the future will unfold, it is certain that it will include change. If times stay good, that is great. That is what we are all praying for. If times get tough, a little insurance is always nice. A form of insurance is the ability to provide for yourself and your family—having the knowledge to produce and preserve your own food is an investment vehicle in its purest form.

Victory Seed Company[1]

Peanuts do not grow on trees, nor do pineapples. Green bell peppers are un-derripe red bell peppers. Parsnips are not related to parsley, but they are re-lated to carrots. Not all potatoes come from Idaho. Ask a nine-year-old where an apple comes from, and he or she will likely respond, "the supermarket."

Americans today are largely food illiterate. For an agrarian nation founded on bounty and diversity, both in agriculture as well as social culture, it is fright-ening how ignorant of food and cuisine we have become. Our conventional sys-tem of agriculture has given us convenience, from the pre-prepared frozen and fast-food meals (so that we do not have to learn how to cook food ourselves) to year-round produce from all over the world (so that we do not have to worry about what grows in our particular region or during what season). Before journalists like Eric Schlosser and Michael Pollan delved into the netherworld of our food-supply system, many folks in America hadn't given much thought to when and from where their food originates, who is growing it and by what methods, and how it is processed and delivered to our food outlets. After a half century of convenience from the industrialization of agriculture, families have cooked less, grown less, but consumed more food—or, as Michael Pollan says, "foodish products . . . that your ancestors simply wouldn't recognize as food."[2]

Much of this forgotten knowledge surrounding food has much to do about supermarket convenience and the amount of processed and fast food in our diet. "Every month more than 90 percent of the children in the United States eat at McDonald's,"[3] notes Eric Schlosser, rendering America a fast-food nation. That such a high percentage of our nation's kids eat at McDonald's is testament to the omnipresence of fast food in our communities and the importance of it in our lives. People—especially children—pick up cues and cravings based on what they see. When we see aisles upon aisles of soft drinks, snack foods, frozen meals, and canned vegetables in our supermarkets, and burger joints, pizza parlors, and drive-through taco stands throughout our community, we quickly forget what was once the most common of common knowledge: what food is, and where it comes from.

As social creatures, we learn from others and from our environment, and what we learn is what we see. This method of learning is what helps define culture: knowledge and traditions about us and our environment that are passed down from generation to generation. Lately, our ancestors have not passed along the culture of growing food or of preparing and eating what we are able to raise in our particular environment. As such, our culture of food production and consumption is very different from that of previous generations. With the rising cost and instability of our food and the decline in our health, it is time we learn what our ancestors neglected to teach us. If public produce is to be feasible—and successful—in our towns and cities, the American population will have to regain that once-great agrarian knowledge that our ancestors possessed. We will have to learn how to recognize edibles from ornamentals (and remember that some can be both). We will have to learn what parts of a plant are edible, and which are not (rhubarb, for example, has toxic leaves; only the petioles—the red stalks—are edible). We will have to learn growing cycles for our particular geography: generally, when certain produce is available, and, specifically, when it is ripe (apples in autumn, citrus in winter, asparagus in spring, apricots in summer). And we will have to learn (and appreciate) forgotten foods that were once commonly enjoyed, but have disappeared from our diet (amelanchiers, or juneberries, which are richer and sweeter than blueberries, are native to the United States, but good luck finding them in the supermarket).

Before we can incorporate better food into our communities, we have to incorporate better food into our vocabulary. Better food choices need to be taught throughout our communities, and public space could become educational in this regard. The trees, shrubs, grass-like plants, and rambling vines that are found along our streets and throughout our parks, plazas, city squares, and the landscaped grounds around our civic institutions can provide—and teach us about—healthy food choices. To be able to see, and eventually recognize, food in all stages of plant development, all around us, is akin to immersion education for a foreign language. Our new language of public produce could become both the medium and the object of instruction in a nation where few have ever had an opportunity to see produce in its native habitat, much less pluck it from the vine. Food does not have to be eaten to have value. Just being able to see the bounty and diversity of edibles in our environments can be educational and may prompt diversity in our diet, while making us more food fluent.

Food literacy helps guarantee a successful system of public produce. To that end, public officials must take the lead in the educational efforts—from informing citizens about what produce is in season for their community and where to find it, to introducing new or unfamiliar produce items and providing guidelines on how to use and prepare them. Municipal staff should explore methods to provide fresh produce to the community at an affordable cost, while teaching individuals how to grow food for themselves so that everyone can partake in some form of self-sustenance.

Strategies for creating and disseminating information to the masses may be quite similar to those employed during the Victory Garden campaign of World War II. For that effort, the government exploited every media outlet available at the time. Promotional and educational articles were published in local newspapers and popular gardening magazines; audio clips and short films were created and broadcast to the community via radio and television; and mimeograph technology allowed pamphlets, handouts, and bulletins to be widely distributed. The information was abundant and comprehensive, ranging from planting techniques to suggestions on which vegetables provided the greatest nutrition, variety, and utility. Recipes, information on canning and preserving, and other preparation and storage tips were also provided, along with methods to extend the growing season and maximize yields. Indeed, the entire structure of the Victory Garden program was formed around the efficient dissemination of information.[4]

Today, the challenge is much the same, but the conduits for conveying information are far greater. Instructional videos and informational downloads can be offered through the city's web site. Many municipalities have access to local cable television stations, and can create custom television programs tailored for their community. Printed material, such as leaflets and fliers, can be included with monthly sewer and trash bills. Food courses and curricula could be offered through the Parks and Recreation Department, supplementing our children's food education that could be available in school-based and after-school programs. The opportunities for local governments to reach the diverse citizens of their communities today are almost limitless, and their potential should be exploited.

The most effective forms of instruction are also the oldest: learning through seeing and doing. Hands-on demonstrations and physical displays are being coordinated by municipalities that employ staff to train citizens in an array of food-growing skills. The City of Portland, Oregon, for example, has implemented an innovative program called "City Fruit," using demonstration gardens to teach gardening skills and promote the environmental, social, and economic benefits of local food production. One community garden hosts the city's "small fruits" demonstration site, which boasts over thirty varieties of both native and exotic fruits, such as strawberries, blueberries, seaberries, pomegranates, and goumi. Portland's most popular demonstration garden is a fruit-tree orchard, comprised of nineteen varieties of apples, three varieties of pears, and four varieties of Asian pears. Staff use the trees to teach specialized orcharding tech-niques, such as espaliering, pruning, organic pest management, and mulching. The sublime beauty of using public produce to teach gardening skills to the community is that the citizens, through their hands-on education, maintain the edibles, relieving the burden on city staff. The most popular lesson is the harvest, when apples are pressed to make cider, offering a salubrious—and delicious—toast to the end of the growing season.

The City of Davenport, Iowa, through the Vander Veer conservatory, re-cently displayed a popular food-growing technique known as "square foot gar-dening." The premise of square foot gardening is simple, and quite effective for urbanites living in upper-level apartments and condominiums who likely have little more than a balcony for growing food. Square foot gardening utilizes raised-bed, intensive cultivation to maximize space and efficiency. Typically, a raised bed measuring four feet by four feet is constructed and filled with com-mercially available garden soil (or equal parts compost, vermiculite, and peat moss). The bed is then divided into one-foot squares, yielding sixteen individual garden plots, each housing its own crop. Though these plots seem miniscule, they are actually an appropriate size for the average family. When we plant a backyard garden, for example, we tend to consume a lot of space constructing rows, and we often plant more than our family can eat. Square foot gardening forces families to garden efficiently and plant only what they can likely con-sume. All sorts of plants can fit within one square foot—from lettuces and greens to broccoli, cauliflower and other cruciferous vegetables, herbs, compact varieties of peppers, eggplants, even a single okra plant. When the produce is consumed and there is nothing left to harvest, the plant is pulled and a new crop is planted. Locating one side of the raised bed next to a balcony railing or fence provides even more gardening opportunities. Tomatoes, peas, pole beans, even squash, can be trained along the railing or fence, encouraging the plants to grow up rather than out. This unique system of gardening can be hard to visualize, but the City of Davenport's display helped people understand the sheer abun-dance and diversity of food that can be grown in tight spaces.

As municipalities begin seeding their public spaces with food, the public needs to be informed as to where the food can be found, how to recognize the

This City of Davenport display teaches Vander Veer park visitors about square foot gardening as well as the biology of pollination.

different produce items, and when they are ready to harvest. Municipalities would do well to emulate the marketing efforts of various food advocacy groups, to not only educate people about available food choices, but provoke interest in healthier eating. With regard to food marketing, there may be no better mentor than the supermarket. Much of a supermarket's overhead is spent on advertising, aimed to elicit excitement from people about various food items for sale in the store. While supermarkets often promote a variety of processed and snack foods, they are well-versed in fresh-produce promotion as well. In the produce aisle of a popular supermarket in Chapel Hill, North Carolina, colorful displays educate shoppers about the varieties of certain produce. One particularly eye-catching display teaches consumers about the different types of citrus. Not only do shoppers learn where various citrus fruits originate, but they learn how to distinguish between them. Within a few moments, shoppers can easily identify a kumquat, a pomelo, and a tangelo—and they learn how to select and

This attractive and informative display helps shoppers in Chapel Hill, North Carolina, recognize, select, and store assorted varieties of citrus.

store the fruits. This sort of education has immense value to the consumer, as it encourages diversity in their diet. Fear is the brood of ignorance, and people are generally apprehensive of the unfamiliar. A little education goes a long way to encourage acceptance of the once unknown, and such a food display helps pique curiosity and develop an appreciation for diverse fruits and vegetables.

Municipalities might also want to take a page from the lesson book of the Center for Urban Education about Sustainable Agriculture (CUESA). Based in San Francisco, this nonprofit organization focuses on food literacy through various programs and celebratory events, while extolling the virtues of eating locally and thus eating seasonally. Some of CUESA's most useful educational guides are their seasonality charts, specifically the fruit, nut, and vegetable calendars. Bay Area consumers can find out just when those prized California artichokes, pistachios, or fava beans are available at the Ferry Plaza farmers' market by consulting the CUESA web site.[5] These calendars also list a variety of unusual

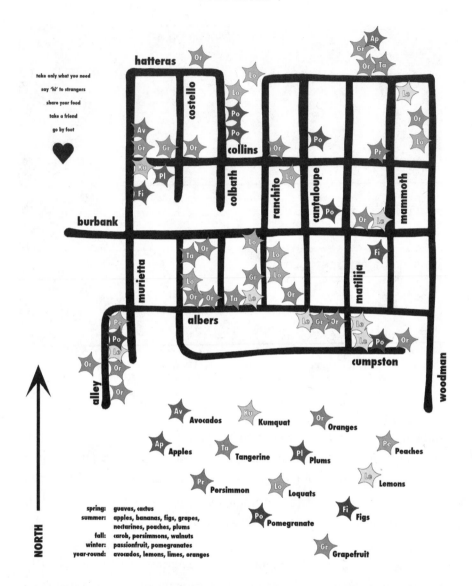

FALLEN FRUIT OF SHERMAN OAKS

this map is a template for free use.   there is no copyright.   learn your fruits!

Map of the Sherman Oaks neighborhood of Los Angeles, locating the different fruit trees with publicly accessible fruit. Each map provides a seasonality legend, coding the fruit to the season in which they are ready for harvest. Each map bears the mantra: "take only what you need, say 'hi' to strangers, share your food, take a friend, go by foot."

crops that reflect both the ethnic diversity and the adventurous foodie spirit that prevails in San Francisco. Produce such as cactus pads and pears, cardoons, burdock, salsify, tayberries, jujubes, and cherimoyas are listed among the more familiar avocados, fennel, figs, bok choy, shallots, and tomatillos. Of particular interest is CUESA's "Seasonal Produce Highlights"—a sidebar on its web site that lists the fruits, nuts, and vegetables that will be at their peak of freshness for the current week. In mid-February, for example, CUESA informs consumers they can expect to see turnips, satsuma mandarins, torpedo onions, fava beans, walnuts, grapefruit, and Meyer lemons. These seasonality charts help consumers better understand their agricultural geography, by teaching them what grows where when. Of course, California is unique, and, because of its mild climate, a tantalizing array of food can be grown throughout the year. But it is just as important for North Dakotans to understand what grows in their neck of the woods, and when it is ripe, so that they, too, can cultivate an interest in food and appreciation for the unique environment in which they live.

Fallen Fruit, the organization described in Chapter 3 that promotes foraging for public fruit in Los Angeles, not only helps inform folks as to which fruits are in season, but publishes neighborhood maps that show exactly where this food can be found. This group has also tapped into Facebook, the extremely popular social networking site. Facebook provides a venue for Fallen Fruit to post its maps and distribute information to an incredibly vast audience; at the same time, other Facebook members can post comments and share their enthusiasm regarding Fallen Fruit's message with millions of people around the world. An increasing number of organizations are taking advantage of the unparalleled outreach opportunity Facebook offers, with its membership now 200 million strong. Municipalities may find it an especially attractive outlet for reaching the young hipsters of the community.

OPENrestaurant, an organization started by a pair of artists in the Bay Area, employs an even more curious, though no less effective, method for increasing food awareness than does Fallen Fruit. According to the group's website, OPEN is a form of "social sculpture"—part restaurant, part art installation, and part performance. More specifically, it says:

OPEN makes available all the physical elements and activities of an ordinary restaurant for artists along with cooks, servers, farmers, activists, community members and educators to develop creative projects that extend the vernacular of food to social, political, economical and environmental issues thus making visible the web of relationships that are brought together in and around a plate of food.[6]

Filling stuff, indeed, and a very novel—and sumptuous—endeavor to educate people about food, where it comes from, and the implications our food choices have on society and the environment.

Most of OPENrestaurant's "performances" take place in cultural centers and art galleries. The concept behind one recent performance was "soil tasting," similar to the notion of *terroir* with regard to wine tasting. Oenophiles understand that terroir—the unique geography, structure, and climate of a particular plot of land, or rather, its "sense of place"—is absorbed in the grapes, and thus, reflected in the wine. Terroir isn't only expressed through wine grapes, however. As OPENrestaurant set out to demonstrate, a soil's unique tilth—its structure, mineral content, and overall health—is reflected in everything it produces. The purpose of the exhibit was to educate people on the subtle nuances of different soils, reinforce their knowledge of unique geographies and food-growing regions, and give them an opportunity to savor the flavors that different soils can impart on the food we eat. For this exhibit, staff grew rocket in diverse soils. Patrons were able to see, smell, and feel the unique characteristics of each soil, and those characteristics were clearly evident in the taste of the tender rocket leaves. Another performance piece, dubbed OPENcity, set out to prepare an entire meal—including bread—from food grown and foraged entirely within the tri-city area of San Francisco, Oakland, and Berkeley. Rabbits came from a West Oakland resident. Salt was harvested from the ocean. Fruit and vegetables were foraged from people's backyards. Even wheat was foraged, but as one participant noted, "it is hard to find wheat in the city."[7] The demonstration was provocative, as was the discussion around the dinner table that night, which sought to answer the question: "How do we make the urban landscape more productive?"[8]

As we work toward becoming more food literate—knowing where our food comes from, who is growing it and how, when it is ready to harvest, and why this knowledge matters—we will also have to expand our food vocabulary to grasp "what is food?" This knowledge is especially useful in helping us to recognize bountiful food items in our environment that we often overlook, largely because we perceive them as ornamental instead of edible, or worse, as weeds. Blackberries and raspberries, for example, are rambling, invasive pests that prove frustrating to control for many homeowners. But in the right setting, they provide a welcome snack. Dandelions have been the scourge of suburban homeowners for decades. Yet dandelion greens are a delicious and nutritious addition to any garden salad. Fennel that is often seen growing wild on roadsides provides both savory aroma and flavor to dishes, and the bulbs are excellent roasted and tossed with pasta. Chayote climbs with reckless abandon in the mild-winter states of the country, but yields wholesome gourds revered by many. Nasturtiums spread like wildfire, but both the leaves and flowers provide pretty—and edible—accompaniments to any meal. Many plants have tremendous value as food crops, but because we have perceived them for so long as weeds, we may not recognize them as food at all.

To help members of our community gain an appreciation of the myriad edible plants in our environment that can be found wild, or those that can be cultivated for food, we may have to circulate recipes, similar to what was done in World War II during the Victory Garden campaign. The suburban homeowner that has cursed dandelions for years might have a difficult time believing that those vexing plants are actually prized salad greens. But one taste and skeptics could become believers. To broaden our food vocabulary and our palate, public officials could organize food festivals that showcase the diverse local foods and how to prepare them. Food festivals are becoming an annual delight in many cities already, lifting community spirit while raising food awareness.

An extensive food vocabulary is crucial to ensure food security—both in being able to produce a bounty of food regardless of one's landscape and to preserve cultures and traditions of our many ethnic families. Food security is commonly defined as "access to nutritious, affordable, safe, adequate, and culturally acceptable food on a daily basis."[9] For homogenous countries, like Finland and Japan, "culturally acceptable" is more tightly defined and commonly understood. But in our multicultural melting pot, culturally acceptable food means a greater variety than is offered through the more typical fresh-produce outlets. As Mark Winne notes, "Many [community-supported agriculture groups] and farmers who sell at farmers' markets are responding only to the food preferences of an educated, white clientele. To be inclusive of a more racially and ethnically diverse customer base, farmers—most of whom are white—have to learn how to grow crops preferred by nonwhite customers."[10]

The need for culturally acceptable produce is perhaps one of the greatest reasons why a central government policy on food will ultimately fail the food security test. Centralized policies, regardless of their reform aims, tend to be of a one-size-fits-all mold. Such a policy regarding food would likely erase the unique traditions and customs of our celebrated ethnic diversity expressed through food and cuisine, replacing them with a more homogenous menu of food items. Food literacy has much to do with an understanding of culture and ethnic diversity, and which foods have meaning and value to the diverse racial groups that comprise our communities. The common globe eggplant, for example, is not suitable for the Chinese, who prefer their own, more slender, delicately flavored variety. Likewise, people of Japanese, Indian, Italian, and other cultures prize specific varieties of eggplant, some of which are not readily available in U.S. supermarkets. Thai eggplants, with their green striped color and spherical shape, bear little resemblance to the common globe eggplant, and are perhaps the most difficult to source.

In most instances, culturally acceptable food is not simply a different variety of a more "common" produce item, but foods that white Americans simply would not recognize as food at all. Ginkgo berries, for instance, are revered by Chinese, Korean, and Japanese families. However, in the United States, the prevailing (i.e., non-Asian) sentiment is that the fruit-producing female trees should be avoided at all costs, as the aromas of crushed ginkgo fruits remind many of an unsavory amalgamation of canine feces, rancid butter, and vomit. Certainly not pleasant—and, for this reason, many people have ardently sought to remove female ginkgo trees from city spaces. Yet when prepared in Asian dishes, ginkgo fruit are quite delicious. The solution, a rather simple but effective one, is to harvest the ginkgo fruit before they fall to the ground and are crushed by unappreciative feet. For folks living on the East Coast, where ginkgoes are found in great abundance, it is common to see people of Asian descent—often Chinese women—gathering the ginkgo fruit. Prickly pear cactus is a spiny, robust plant with high ornamental value to white Americans for their water-conserving gardens. But to Mexican families, *nopales* is a staple in their traditional cuisine. Callaloo (amaranth greens) is a much sought-after leafy green vegetable in African-Caribbean communities, though few other cultures have developed a taste for it.

Of course, context is paramount to the success of public produce. Female ginkgo trees may be unwelcome in predominantly white suburbs of America. Callaloo would likely be regarded as a weed. Prickly pear may be more acceptable,

Prickly pear cactus may not look like food to many people, but *nopales* (the Spanish name for the large, flat pads) and prickly pear (the fruits atop the pads) are highly regarded in Mexican communities, and a staple in both authentic Mexican and New Mexican cuisine.

but only as a specimen in a xeriscape garden. But when certain fruits that are deemed unpalatable to others are planted where citizens not only recognize them as food, but prize them because they are part of their rich cultural heritage—and because they cannot be found through the typical food outlets—they add immense value to public space, and to the city as a whole.

We also need to recognize that ethnic food that was at one time "foreign" has now become (or is capable of becoming) culturally acceptable to the masses, including white America. Not too long ago in California, for example, bok choy and daikon radish were virtually unheard of except in Asian markets, and could have been deemed, at the time, culturally unacceptable to many Americans. During the late 1980s, however, the popularity of Asian vegetables skyrocketed—almost overnight. Between 1988 and 1989, the production of Asian vegetables in California increased 41 percent.[11] Today, bok choy and daikon radish are readily found in supermarkets throughout the state (indeed, throughout the country!), enjoyed by Chinese, Hispanics, blacks, and whites. Chinese cuisine has become immensely popular in America, as has Japanese, Indian, Mexican, Ethiopian, Thai, Vietnamese, Caribbean, and Greek, illustrating that foods that are still typically regarded as culturally "foreign" can also be highly desirable.

We need to realize that in the United States, where diversity is celebrated, culturally acceptable foods open a world of gustatory opportunity, while addressing issues of food security. For this very reason, city staff in Des Moines purposely plant produce varieties that are quite unfamiliar to Iowans. Some examples found throughout the city's many edible landscapes include jostaberry, pawpaw, medlar, Asian pear, Chinese chestnut, and juneberry. As Teva Dawson of the city's Parks and Recreation Department noted, "It was very important for us to choose some varieties that we knew were hardy and disease resistant as well as more unique and unusual plant varieties that folks would not find in their grocery store."[12] Obviously we need to ensure apples, peaches, pears, tomatoes, lettuce, carrots, potatoes, onions, and other recognizable staples are provided within our edible landscapes. But we also need to become more food fluent, and what is necessary are programs that teach the public how to harvest, prepare, and eat foods such as loquat, carob, sorrel, nasturtium, fennel, dandelion, chayote, prickly pear, jujube, passion fruit, pineapple guava, and a host of other vegetables and fruits that our ethnic communities (and a growing number of restaurateurs, for that matter) across the nation prize. In doing so, we will reinforce that, in America, such "unusual" foods are, indeed, culturally acceptable.

As we accept a more varied menu of food choices, landscape architects, as the principal designers of urban open space, will have to sharpen their agricultural acumen as well. Seasoned landscape architects have committed to memory hundreds of trees, shrubs, and groundcovers—their botanical names as well as their common ones. The vast majority of their plant knowledge, however, is lopsidedly focused on ornamentals. Currently, edible landscapes are a niche market for the few landscape designers who wish to take it on. In the future, it will no longer be a niche. Landscape architects will need to be versed in ornamental *and* edible plants, and how to plant them in combinations that not only create beautiful compositions, but compositions that realize the principles of integrated pest management and companion planting, while providing utility, joy, comfort, and relief to people in public space.

Perhaps the most important social group that benefits from food education is our children. Food literacy, like language, is most effective when it is taught at a young age, and many experts say that food and dietary choices taught early in life set lifelong patterns. For these reasons, many communities are starting to incorporate schoolyard gardens as part of the education curriculum. Schoolyard gardens as public education have a long history in this country, dating back to 1890 with the Putnam School in Boston.[13] Since then, their popularity has waxed and waned with the changing agricultural attitudes and economic times. Today, they are enjoying a renewed vigor, which could not be more timely, given the rise in childhood obesity and the preponderance in our children's diets of processed and fast foods—many of which, ironically, are available to kids in their school cafeterias.

Schoolyard gardens teach children basic life lessons, such as self-sufficiency, the natural cycles of life, and, more rudimentary, where food comes from. Many children today have no idea that potatoes come from the ground, nuts from the tree, or grapes from the vine. Teaching kids to garden gives them a sense of accomplishment, while exposing them to a diversity of whole foods and the miracle of Mother Nature. Most importantly, teaching children about the benefits of whole foods establishes a pattern of healthy eating, and changes our food culture from fast food to slow food. Michael Pollan contends that we may never fully develop an appreciation and fondness for healthy foods until there is a significant change in our food culture. And that change, he says, "must begin with our children, and it must begin in our schools." He further maintains that eating well is a critical life skill, and that "we need to teach all primary-school students the basics of growing and cooking food." Much like President John F. Kennedy recognized the woeful physical ability of our nation's children a half century ago, and required physical education to be established in public schools, many today, like Pollan, are arguing for an equally important "edible education."[14]

Perhaps the most famous—and comprehensive—edible education program in the nation is found at Martin Luther King Jr. Middle School in Berkeley. The Edible Schoolyard, the brainchild of culinary maestro Alice Waters, is a one-acre organic garden and kitchen classroom. Before the garden was created, the grounds around the school were comprised of asphalt, dried lawn, and leggy perennials—certainly not the landscape that inspires imagination. Today, a veritable oasis of organic produce encourages children, teachers, and parents to learn more about food, nature, and healthy eating.

The mission of the Edible Schoolyard is unique, and aims to "create and sustain an organic garden and landscape that is wholly integrated into the school's curriculum, culture and food program. ESY involves students in all aspects of farming the garden and preparing, serving and eating food as a means of awakening their senses and encouraging awareness and appreciation of the transformative values of nourishment, community and stewardship of the land."[15] After a quick walk through the facility, it is clear that the goals are being met. The garden is planted with seasonal produce, herbs, berries, and fruit trees. There is also a tool shed, seed propagation table, chicken coop, and a pizza oven.

The Edible Schoolyard is part of the curriculum for all students at King Middle School. In the fall, sixth graders—as part of their math and science curriculum—work in the garden while seventh graders in the humanities and social science classes start in the kitchen. Come spring, the classes trade places. The eighth graders use the garden year round for science classes or specialized projects. When finished with the curriculum, the students will have gained a complete seed-to-table experience, which begins by preparing and seeding the planting beds and concludes with a sit-down meal at the table, complete with flowers from the garden. And, yes, the children even participate in the cleanup. Throughout the school year, the program exposes children to food production, ecology, nutrition, and fosters an appreciation of meaningful work, and of fresh and natural food.

The Edible Schoolyard program at King Middle School has become so admired that the garden is visited by more than a thousand people each year, from all over the world. Educators, health-care professionals—even legislators—come to learn how the seeds of this program could be planted in their communities. This unique program has spawned hundreds of kitchen and garden programs throughout the United States, and official Edible Schoolyard affiliates have been established in San Francisco, New Orleans, and Greensboro, North Carolina. Because of its comprehensive curriculum and the important life lessons that are taught, the Edible Schoolyard is arguably the benchmark by which any food-education program is measured, whether it is aimed at children or adults.

Food education, as Pollan inferred, should be compulsory within all primary school curricula. But such a course has value even to institutions of higher learning. Harvard University aims to graduate more food-literate students through its Food Literacy Project. A partnership between Harvard's Dining Services, the School of Public Health, and the school's Health Services, the Food Literacy Project "cultivates an understanding of food from the ground up." Specifically, the program aims to promote an enduring appreciation of "four integrated areas of food and society: agriculture, nutrition, food preparation and community." Harvard's program offers cooking classes to the "culinary challenged," helps students decipher supermarket food labels and culinary jargon on restaurant menus, and has compiled a library of some of the most provocative book titles on the subject of food, providing satiety to both the body and mind. Most importantly, the Food Literacy Project teaches students about farming practices and where food originates, reminding them "agriculture is at the heart of every meal."[16]

What students and teachers have invariably learned through food-literacy programs is what others across the country will soon discover as we embark on the needed changes in our food production system: Local, organically grown produce, whether it is grown on rural fields or urban plots, looks very different than what is commonly found on supermarket shelves. For some, organic can be ugly. Ugly, however, does not mean unpalatable. It does mean undesirable to supermarkets, and that is a shame. We have become a very finicky society (some would say superficial) with regard to produce. Supermarkets control the standards of produce, and only the best-looking food finds its way onto store shelves. Produce that is too large or too small, misshapen, discolored, or exhibits a nick, crease, or other blemish that might render the item "unsightly" is discarded. Yet, anyone who has ever tried to grow their own fruits and vegetables understands just how few perfect specimens are produced in nature. Indeed, it is this stringent standard for aesthetic quality that give food banks a good supply of fresh fruits and vegetables.

Marguerite Nowak, an advocacy manager for the San Francisco Food Bank, says that the majority of the 16 million pounds of produce that her organization distributes annually comes from farmers' stock that fail to meet supermarkets' standards of beauty. "When we head down to the fields, we get a lot of oranges, for example, that are not the right size or color for the supermarket. But they are fine otherwise and taste just as good," she observes.[17] Only the finest specimens are deemed saleable by the supermarkets, and because they have been the most common source of our produce throughout our lives, we might think a forked carrot or a heart-shaped potato is somehow diseased. But malformed carrots and potatoes are part of any agriculture, rural or urban. Ditto for worms and bugs in spinach, lettuce, and other greens. Tomatoes with minor splits in the skin or occasional brown spots are probably more wholesome and delicious than their pretty sisters sitting on the supermarket shelf. We should be judging food not by its form and complexion, but by its provenance.

As public officials provide edibles in our public landscapes, and we encourage folks to forage and plant their own, we have to educate people on the rules of public-space cultivation and harvest. The orchard established in the neighborhood park means the fruit is available to *all* members of the public, whether they live in the neighborhood or not. It is natural for the people most responsible for maintenance and care of the food to assume some degree of ownership, and to want to control who the produce is distributed to. While it is reasonable to expect some restriction on who can harvest (say, ousting the wily entrepreneur who harvests all the apples in the dark of night and sells them back to the community at the farmers' market the next morning), the reality is that, by legal right, the produce is entitled to anyone and everyone. Likewise for the edibles that many homeowners plant in the public right-of-way along their sidewalk or in the street median. Residents often consider these spaces to be an extension of their private front yard gardens. Though the difference in public versus private produce may be subtle to the pedestrian, the property line is distinct and the legal definition clear. Food grown in (or, in the case of Los Angeles, that is accessible from) the public right-of-way belongs to the public. One who takes the time to plant, grow, and tend the fruit tree in the space between the sidewalk and curb is not granted ownership. Rather, his or her efforts should be regarded as a selfless, charitable contribution to the community. While it would be ideal if the passerby always asked permission before grabbing some apples or plums, the homeowner should not expect it. With regard to fruit trees, however, such harvests from strangers are not entirely unwelcome. In my experience, folks who plant fruit trees usually find they have more food than can feed their family. Indeed, many homeowners encourage others to harvest, as it saves them from the burden of harvesting (or disposing of) everything themselves. The simple message municipalities should be conveying to the community and its foragers could echo the mantra printed on Fallen Fruit's public fruit maps: "take only what you need, say 'hi' to strangers, share your food, take a friend, go by foot."[18]

Municipalities should encourage, rather than forbid, home and business owners to plant edibles in the right-of-way. It should be made clear, though, that it is the merchant's or resident's responsibility to care for the edibles and maintain a clean and clear public sidewalk. Such knowledge of local ordinances helps a network of public produce succeed with minimal disruption and contention. When everyone is aware of his or her rights and responsibilities, and understands how planting edibles benefits both the individual and the group, public produce can help promote a culture of sharing in the community.

The myriad programs, events, demonstrations, and displays created by food advocacy organizations around the country illustrate the creative strategies to increase food literacy. The benefits of being food fluent are that we gain an appreciation for food, its diversity, and all the places and methods of its production. These educational efforts foster nutritional, social, and environmental awareness, and are worth emulating by city planners and public officials. Municipal staff would be wise to create a custom-tailored lesson plan by gleaning those techniques and messages that will provide their citizenry the greatest value. Once the information is assembled and disseminated, public officials can be assured of a healthier, food-literate community.

CONCLUSION

Community Health and Prosperity

It is more difficult in this country and these times to find a patch of land to scatter seed on, and people to work it in the long term. In red-lined neighborhoods vacant lots are there for the taking, often full of toxic waste. In many areas, the younger generation lost traditions of cultivating the land. In some cities gentrification has turned the land into a commodity far out of the reach of gardeners, where people have to work so much to pay rent that there's not time to produce food in a serious way. In others, car culture has so isolated people that they don't even know their neighbors.

But everyone's got to eat. And while gardens aren't a cure-all to the problems of economic racism and environmental injustice, unequal access to resources and an exploitative profit system, they can help us get by a little easier, give us space to breathe, to learn from the earth, and to begin to reweave relationships based on respect for the land and for the people around us.

Cleo Woelfle-Erskine, *Urban Wilds*[1]

Public space permeates cities, giving them order and form, and is arguably every city's greatest, most extensive physical resource. The numerous streets, parks, squares, and plazas are where people from all walks of life come together to socialize with familiar faces, make new acquaintances, and simply revel in one another's company. Public space thus builds community, and because this is where different people share different ideas, public space also helps sow the seeds of democracy. In these ways, public space nourishes our soul. But might these places nourish our body as well?

Growing food in public space is not a new idea, but one that is timely and worth revisiting. From the late-nineteenth century to the mid-twentieth, government had asked city dwellers to help mitigate economic and social distress through urban agriculture on public land. These urban agriculture endeavors sought to grow food on every available and arable plot of land in our cities, private as well as public, to supplement a declining food supply, and help establish food security. The efforts from the Victory Garden campaign, arguably the most successful of the different government sponsored urban agriculture efforts, were staggering. By 1944, there were an estimated 20 million victory gardens yielding eight million tons of food, collectively providing 40 percent of the nation's vegetable supply.[2]

More than food, however, victory gardens and their earlier urban farming counterparts promoted self-reliance, self-respect, and economic independence, and provided financial, physical, and spiritual well-being. And because these forms of public produce were established in many neighborhoods, involving many neighbors, they helped build and nurture community, as well. Some cities, like Boston, have never lost sight of the value of these earlier programs of urban food production. Indeed, the seven-acre Fenway Victory Gardens, located along the Muddy River in the Back Bay Fens, is the oldest continuously cultivated victory garden in the country, flourishing since 1943. Other cities, like San Francisco, are rediscovering the value of victory gardens. In 2008, the City of San Francisco funded a Victory Garden pilot program that involved fifteen households representing a diverse cross-section of the community and, most notable, established a seasonal, edible garden in front of City Hall. Seeking to increase local food security and decrease the food miles associated with the average American meal, San Francisco's program equates "victory" to fewer carbon dioxide emissions, self-reliance, seasonal growing and eating, community action, and, most notably, independence from corporate food systems.[3]

The goals of San Francisco's Victory Garden program illustrate the broad goals of community food production, and the inherent environmental and societal benefits of growing food on urban land. In fact, the goals of the original Victory Garden campaign of World War II were also broad, affecting not just food security, but individual economic assistance as well as family morale. During a National Defense Gardening Conference, held on December 19, 1941, the secretary of agriculture and the director of the U.S. Office of Defense, Health, and Welfare Services articulated the goals of the Victory Garden campaign:

> Increase the production and consumption of fresh vegetables and fruits by more and better home, school, and community gardens, to the end that we become a stronger and healthier nation.
>
> Encourage the proper storage and preservation of the surplus from such gardens for distribution and use by families producing it, local school lunches, welfare agencies, and for local emergency food needs.
>
> Enable families and institutions to save on the cost of vegetables and to apply this saving to other necessary foods which must be purchased.
>
> Provide, through the medium of community gardens, an opportunity for gardening by urban dwellers and others who lack suitable home garden facilities.
>
> Maintain and improve the morale and spiritual well-being of the individual, family and nation.[4]

The goals for victory gardens then are just as relevant as those of public produce today. There are some differences in organization and operation, however. For one, the big push for urban food security during the Victory Garden campaign came from Washington. Whereas the federal government provided encouragement and guidance, municipal government, and its local citizens, provided action. For future urban food-producing endeavors, municipal government will likely have to provide both impetus and implementation. Certainly, if central government today adopted a food attitude similar to that exhibited during World War II, it would be welcome. In all likelihood, however, today's federal government, with its now sixty-year favor of centralized, industrial agriculture, seems unlikely to change. And even if that attitude does change, it may not be as effective as local-government policy. As Mark Winne argues, "Democracy works best when it's closest to the people. That is why we can expect city hall to act faster than the state capitol, which in turn tends to respond to its people before Washington D.C. The farther away the decision makers are from those whose lives are affected by their decisions, the slower will be the change that occurs."[5]

Public produce will also need to be regarded with more permanence than the earlier food security programs. The Potato Patches, Liberty Gardens, Depression Relief Gardens, and Victory Gardens were all ephemeral (Boston's Fenway Victory Gardens excluded). They existed within a finite time of economic and social distress, and promptly vanished once prosperity rebounded. Though the success of victory gardens provoked many to fight for their continuance, the end of the war and the transition to an industrial system of agriculture left the government—and the general public—without desire to maintain the prolific and abundant urban gardens. While the exploding middle class and higher socioeconomic strata enjoyed benefits from the industrial agriculture boom, the inner-city poor were still left without an adequate supply of food. As suburbs consumed farmland outside the cities, and supermarkets and grocery stores followed the mass emigration of the post-war population, food problems were exacerbated for inner-city residents. Today, however, food insecurity is threatening more than the urban poor. Folks across all socioeconomic standings are feeling those hunger pangs that our grandparents and great-grandparents felt early last century. And this time, even if our current conflicts in Iraq and Afghanistan are resolved, and we see fiscal stability yet again, the continued depletion of our finite oil supply and the irreversible trend of global warming will require community food-producing activities with longevity. History has taught us that we can no longer believe that community health and prosperity lies in a centralized, corporate system of food production. Indeed, it is precisely this system that has contributed to the food crisis we currently face. If allowed to continue, and if history of other fallen nations is any indicator, we may lose any chance of regaining economic prosperity.

Eric Schlosser, in his best-selling book *Fast Food Nation*, drew frightening parallels between our current system of agriculture and that of the former Soviet Union:

> Throughout the Cold War, America's decentralized system of agriculture, relying upon millions of independent producers, was depicted as the most productive system in the world, as proof of capitalism's inherent superiority. The perennial crop failures in the Soviet Union were attributed to a highly centralized system run by distant bureaucrats. Today the handful of agribusiness firms that dominate American food production are championing another centralized system of production, one in which livestock and farmland are viewed purely as commodities, farmers are reduced to the status of employees, and crop decisions are made by executives far away from the fields.[6]

Recently, this nation has witnessed the vulnerability of such a centralized system. When storms flooded the corn fields of Iowa in June 2008, inundating 1.3 million acres of that cropland, amidst rising fuel prices and the divergence of corn from food to biofuel, prices for beef, pork, poultry, eggs, milk, and cheese soared not only across the country, but throughout the world. Our reliance on oil to produce corn in Iowa, and corn to produce ethanol to combat the high price of oil, and the weather anomalies and climate change that have resulted from decades of burning both oil and ethanol, has locked us into an unsustainable cyclone that, if left to perpetuate, will decimate our food supply. The complex interplay between corn, oil, climate change, and the price and availability of food today illustrates an urgency to return to a more environmentally sustainable, decentralized system of agriculture, closer to—and within—our cities, where the majority of people are currently living, and will likely continue to do so. A network of public produce, grown on underutilized public land using plentiful and readily available sunshine, can help decentralize our system of agriculture and increase food security.

Critics, I expect, will contend that serious food production requires land—lots of land—and the skill, knowledge, and machinery necessary to produce the quantities of food citizens demand. The Victory Garden effort, history has taught, proves otherwise. Regardless, public produce is not meant to annihilate the current food machine, or undermine the relationships that communities are building with smaller, regional farms. My hunch is that there will always be a need for some form of corporate industrial agriculture (though it will have to be scaled back), and there will always be a desire to have independent family farms within a couple hours drive of the city. Merely, public produce offers a city's citizenry yet another choice, and there is certainly room for all. That is what cities have always been, and should continue to be, about: providing choice. But the fact remains that the vast majority of agricultural land does not produce food to feed us directly. Rather, the overwhelming majority of agricultural land is used to produce feed for livestock, or else it is used in the manufacturing of processed foods, such as high-fructose corn syrup, partially hydrogenated fats, and enriched grain flours that find their way into the myriad soft drinks, cereals, snack foods, and baked goods that currently represent staples in the American diet.

Farmers' markets have recently exploded in popularity and provide healthful, much-needed dietary alternatives to the big food-producing corporations. While the quality of what is offered at farmers' markets is often better than what is found at supermarkets, this lower-yield, higher-quality comes at a price. Community-supported agriculture groups can offer a more cost effective alternative, but one is more limited in choice. A member of any such group only receives produce that his or her CSA grows. This is especially disappointing for ethnic families, who seek the more "exotic" produce items for their unique cuisine. And CSAs generally do not give you the opportunity to specify how much of a particular produce item you want (or may need) each week; you get what you get. While CSAs and farmers' markets can add variety and help supplement what we buy at the grocery store, there is still need for additional, cost-effective produce choices, especially for those who cannot afford the alternatives. This is where municipalities can help.

As civil servants, municipal officials need to implement programs and policies that meet the needs and desires of their citizens, and help in the implementation of food producing activities. Food security has been an issue for many urban communities for decades, and, over the years, there have been many attempts at urban food production. Usually founded on the premise of "Golly, wouldn't it be great if everybody grew their own food and there would no longer be issues with hunger and food security?", these attempts ultimately fail, principally because of the lack of initiative and support from the community leaders and policy makers.

During the 1996 American Community Gardening Association conference, Laura Lawson brought up the intractable concerns of sustaining urban gardening longevity. In her book *City Bountiful,* Lawson writes "Once we put aside our pictures of happy children and inspirational stories, we found that we were all struggling with similar concerns. Everyone was committed to the idea of gardening as a resource to serve the social, environmental, and economic needs of urban, low-income communities, but everyone also felt pressured by insecure land tenure, competitive funding, staff burnout, and the need to sustain community-based leadership."[7] These issues, which have persisted for decades with community gardens, will likely continue, unless local government adopts a proactive attitude and takes a more hands-on approach to fresh-produce production. Securing land, allocating funds, and dedicating municipal staff to the issues of public produce can help ensure community food security. And it goes beyond meeting the food needs of the poor. In addition to assisting low-income communities that currently have limited access to fresh produce, public produce can aid middle-class Americans, who are increasingly discovering that farmers' market produce is beyond their financial reach. As well, it can meet the demands of the upper class, who increasingly insist that the food they consume—be it in their homes or in posh restaurants—be locally grown. And public produce can aid all folks who have simply grown tired of conventional agriculture and desire a little glimpse of the agrarian life that this land of opportunity was founded upon.

Aside from obvious economic and environmental benefits of local food production efforts, there are intrinsic civic benefits associated with public produce. Hopefully, municipal government, as it works toward that ultimate goal of building healthful and prosperous communities, will embrace the social benefits of public produce as being equally important as the environmental and economic ones. As Luc Mougeot contends, "Urban agriculture is not the total solution to issues facing the future of cities, but it is an essential part of any program to make those cities more livable, and to improve the lives of city dwellers."[8] One societal characteristic of urban agriculture that differs from conventional agriculture is the democratization of community-based food production. Unless food consumers are part of the family corporations of industrial agriculture, or sit on the boards of the larger, public corporations, they have little voice in food choice. Voting with one's pocketbook is not entirely rational when it comes to food, as it often is with other retail endeavors. People have to eat. If the choice is between supermarket produce that they may be able to afford versus farmers' market produce, which is certainly out of their financial reach, there is truly no choice. Such purchasing patterns erroneously suggest that people "prefer" supermarket produce, because that is what the majority of people buy.

Public produce, on the other hand, gives consumers a significant voice in food production. Because it is on public land, the public has a direct say in how that land is managed, either through the democratic process of electing leaders (mayors and members of the council), lobbying those already elected about food policies, or merely relating concerns to the stewards of those lands (municipal staff members or community volunteers, for example)—those they see every day on their way to work, school, places of commerce, recreation, or worship. And, not to forget, these municipal land stewards—city staff that grow the produce, and maintain the garden plots—are themselves members of the community.

A system of public produce thus yields a new type of member in the community—one the late Thomas Lyson called a "food citizen." The idea is that public produce, a form of food production that fits within Lyson's definition of "civic agriculture," gives citizens a voice that is currently muted in our conventional food-supply system. Lyson reasoned:

Civic agriculture flourishes in a democratic environment. Problem solving around the social, economic, and environmental issues related to agriculture and food requires that all citizens have a say in how the agriculture and food system is organized. Indeed, citizen participation in agriculture and food-related organizations and associations is a cornerstone of civic agriculture. Through active engagement in the food system, civic agriculture has the potential to transform individuals from passive consumers into active food citizens. A food citizen is someone who has not only a stake but also a voice in how and where his or her food is produced, processed, and sold.

The free-market neoclassical system of conventional agriculture, on the other hand, does not necessarily benefit from democracy and, in fact, may be constrained by the politics put into place through democratic actions of citizens.[9]

Civic agriculture, as Lyson contended, holds value to everyone in the community. One group that society increasingly finds difficult to engage is teenagers. After Mark Winne started the Natick Community Farm in Massachusetts in 1975, he and his colleagues had found that the farm had an unexpected value for teenagers, especially at-risk youth. Principally, community farming, as Winne discovered, provides a positive outlook for these youngsters. He noted, "It has . . . become increasingly important that the town's young people have an alternative frame of reference that doesn't include the local mall and that gives them a respite from an economic system that treats them as if they are only consumers-in-training." The ultimate value, Winne realized, was that these urban agriculture endeavors "have amply demonstrated that life offers a richer menu of choices."[10]

If San Francisco is any indication, this richer menu of choice even has value for society's seemingly incorrigible. Similar to Winne's observations that urban agriculture gives teenagers an alternative frame of reference that doesn't include the typical teen hangouts, urban agriculture offers offenders in San Francisco an alternative frame of reference from the troubled streets of their community. Former inmates raise organic vegetables, such as radishes, kale, Swiss chard, and broccoli—as well as many varieties of fruits—on land owned by the San Francisco City and County jail. The Garden Project, the moniker of this unique urban-agriculture program, aims to prevent crime and reduce recidivism. And it is proving quite successful. San Francisco County sheriff Mike Hennessey notes, "The Garden Project is a tremendously effective crime-prevention program. It not only helps individuals rebuild their lives, but recidivism studies we've conducted also show that while 55 percent of our prisoners are rearrested within a year, those who go through The Garden Project have a recidivism rate of 24 percent, and that's after two years."[11]

Crack dealers, hookers, assailants, and scores of at-risk youth have gone on the straight and narrow simply from growing organic food. Outside this municipal garden, the same streets are still wracked with crime, homelessness, drugs, and prostitution. But through gardening, these former offenders find solace, and learn about hope. *New York Times* reporter Jane Gross, who covered The Garden Project, poetically relayed the deeper meaning of growing organic produce, describing fruits and vegetables as

> metaphors for what went wrong in a prisoner's troubled past, lessons about how to live a healthy and honorable life, and proof that love and work make a garden flourish. . . . The simple process of weeding is a good place to start re-examining a life gone wrong. The weeds are whatever got in the way: smoking crack or whoring or stealing. Once they are gone, no longer leaching water and nutrients from all that grows around them, the vegetables and fruits thrive.[12]

Mark Winne believes that "the power of community gardening and other similarly organized small-scale farming efforts in nontraditional areas such as urban America is not found so much in the rate of return to the food supply but in the rate of return to society."[13] When criminals are rehabilitated, that rate of return to society could not be any higher. These one-time public offenders, who used to hurt people, now help them. The food that is produced by The Garden Project used to be sold to many tony restaurants in the San Francisco Bay Area. Today, the produce has a higher calling; it is used to feed the city's many elderly and poor families. The one-time criminals growing this food earn a modest income, which helps them get by from day to day. What has greater value to them, and to society, are the life lessons they learn, which yield positive returns beyond their immediate future.

The Garden Project, and any system of public produce, could also provide a manifestation of author-activist Van Jones's notion of a "green-collar" workforce. His idea of a green-collar economy arose from the "Green jobs, not jails" mission of the Ella Baker Center for Human Rights, an organization Jones co-founded in 1996. That organization advocated for reformation of the juvenile justice system through green-collar jobs, giving at-risk youth economic opportunities to help lift them out of poverty. The inspirational ideas and pioneering work has since evolved into Green For All, an organization founded by Jones that creates "entrepreneurial, wealth-building opportunities for those who need new avenues of economic advance."[14] According to Jones, who was recently appointed as "green jobs" adviser to President Obama, the establishment of urban, sustainable gardens is just one of the avenues that Jones contends is necessary to combat global warming, while possibly creating job opportunities for the urban underprivileged. In this way, public produce helps these people turn their lives—and the planet—around.

Imagine the comprehensive network of public produce that could result if one municipality aggregated all of the individual urban agriculture efforts undertaken in other communities: the San Francisco sheriff who helps organize a rehabilitation garden for former offenders in his community; the mayor in Chicago who orders a honey-bee colony on the roof of City Hall; the parking manager in Davenport who grows vegetables outside his office window, and turns an underutilized parking lot into a community garden; the city-owned, vacant parcels of land in Detroit that have been turned into acres of community gardens; the Parks Department employees in Portland and Des Moines who plant fruit trees in neighborhood parks; the planning policy in Seattle's Comprehensive Plan, mandating that at least one community garden be established for every 2,500 households; the Los Angeles organization Fallen Fruit that publishes newsletters and maps alerting people to the whereabouts of publicly accessible fruit in the city, and when it is ready to harvest; and the restaurant staff in Berkeley who are willing to trade meals for Meyer lemons. All of these seemingly disparate local agricultural and food-production efforts, coalesced into one program that provides assistance to the individual while bettering the community, is certainly a model worth creating, and subsequently emulating.

The communal benefits of public produce are many, and more will be unearthed, once public gardens are established. As the City of Chicago's Eat Local Live Healthy program contends, "Urban agriculture provides healthy food, aesthetic improvements, and increased interactions among neighbors in cities across the country."[15] For public produce to be effective, however, government officials, as demonstrated by the City of Chicago, need to be involved. There are a host of obstacles to urban agriculture, not the smallest being disinterest from government across the country: federal, state, as well as local.

A few progressive municipalities have begun clearing those obstacles, now that they recognize that food security and community health are inextricably linked. Municipal staff and elected leaders across the country are realizing that the need to create policies to ensure communal access to healthful, affordable food falls squarely on the shoulders of local government. Rather than try to avoid the issue of urban food-production, as many public agencies have done in the past, it is time to recognize the benefits of local agriculture efforts and take a more proactive approach to a healthier, more prosperous community. More importantly, public agencies need to lead by example. Public produce should be a policy within every municipality, whether that policy lies in the department of Community Planning, Parks and Recreation, Public Works, Forestry, or all of the above. The City of Des Moines has taken great initiative to ensure food security and help establish prosperous communities. That community's efforts are staggering and laudable, spanning the range between education, planning, maintenance, implementation, and community development. In 2004 alone, the Des Moines Department of Parks and Recreation helped establish over seventy community gardens for schools, neighborhood associations, libraries, shelters, and other civic facilities; promoted the "Plant a Row for the Hungry" campaign, urging community gardeners to grow extra produce that could be donated to the local food bank; developed edible landscape designs for Martin Luther King Jr. Elementary School, Orchard Place (a mental health and juvenile justice facility), and the Capitol Park Neighborhood Association; delivered thirty backyard kitchen gardens to low-income households in the Carpenter and Capitol East neighborhoods; provided on-site training and consultation; taught community groups how to use urban gardening as a community development tool; awarded more than thirty grants for gardening supplies and other resources; and held an eight-week training class on organic gardening.[16] Des Moines' success in helping to establish food security for its citizens is the direct result of active involvement from municipal staff with support from city leaders.

There has been a recent call by some of our greatest food advocates and thinkers—like Eric Schlosser, Marion Nestle, Michael Pollan, and Alice Waters—for a change in federal government policy toward food and food production. The first step toward policy reform, many argue, must come from the American president himself. Michael Pollan contends there is tremendous "power of the example you set in the White House. If what's needed is a change of culture in America's thinking about food, then how America's first household organizes its eating will set the national tone, focusing the light of public attention on the issue and communicating a simple set of values that guide Americans toward sun-based foods and away from eating oil."[17] Such a change in mindset is timely and prudent. A simple, symbolic gesture, one originally proposed by Alice Waters during President Clinton's administration is to convert a portion of the South Lawn at the White House into an organic vegetable garden. Though this may sound unconscionable to formalist architects and landscape designers, the idea of a fussy vegetable garden on the verdant, immaculately manicured lawn isn't new. In fact, during World War II, Eleanor Roosevelt used a sizeable portion of the White House lawn to create a victory garden, inspiring Americans nationwide to do the same in their own backyards. Such a conversion today, from a sterile, purely ornamental landscape to an edible garden of bounty, would certainly be a signal that our national leader cares about what people eat, and that changes in the manner our food is produced and distributed are necessary.

Waters's suggestion fell on deaf ears during the sixteen years of the Clinton and Bush administrations. Finally, in March 2009, the new First Family heeded Waters's advice, and ground was broken for a 1,100-square-foot organic produce garden that would be planted with dozens of different vegetables and fruits. The principal reason for the garden, according to First Lady Michelle Obama, was "to educate children about healthful, locally grown fruit and vegetables at a time when obesity and diabetes have become a national concern."[18] The Obamas are confident the modest garden will send a symbolic message to families throughout the country that it is time to reassess our current eating patterns.

Concurrent to the announcement of the White House veggie-garden, President Obama announced new appointments to the FDA along with tougher food safety measures. Obama recognized that the FDA has long been understaffed and underfunded. He cited the *E. coli* outbreak in 2006 and the *Salmonella* outbreaks in 2008 and 2009 as principal reasons stricter food regulations need to be in place. During a weekly address, the president remarked, "These incidents reflect a troubling trend that's seen the average number of outbreaks from contaminated produce and other foods grow to nearly 350 a year—up from 100 a year in the early 1990s."[19]

While the White House vegetable garden is inspiring, and Obama's plan to bolster FDA staff and inspections long overdue, the root of the problem is still the elephant on the dinner table: our corporately-controlled, centralized system of agriculture. Obama's reinforcements to the FDA are needed to ensure the foods we consume from packers, processors, and distributors are safe. And the veggie-garden sends a loud and laudable message to families who can afford both time and money to grow their own food. But these initiatives don't do much to modify behavior of the current beast, by rethinking how we can deliver wholesome nutritious food to citizens who need it most—those who do not have the time, ability, or financial wherewithal to secure it for themselves. One point that merits criticism from the First Garden is that the produce is not intended for any other persons other than the Obamas (and their dinner guests). The next evolution of urban agriculture will have to focus on fresh, healthful, organic, local food of the people, by the people, and for the people, to paraphrase Abraham Lincoln. The First Family's garden is an ennobling gesture, but urban agriculture needs to do more by serving more.

The measures that President Obama has taken to ensure a safer and healthier food system in our country are praiseworthy, and hopefully, a presage. At this time, there has not been any firm indication that he is considering an overhaul to our current, industrialized system of food production and distribution (other than ensuring that the foods we currently eat are safe from pathogens). Even if that change in mindset does come, I am less confident that a centralized policy for food will do much to change our current centralized system of agriculture. It is difficult to ensure that broccoli, for example—a crop that can be grown in almost every state, but which more than 90 percent is harvested from California[20]—reaches each and every neighborhood in America. So, too, would it be difficult today for food policy from Capitol Hill to filter through to every Main Street in America. Food is more than simple sustenance; it is an indicator of culture and a binder of community. Our communities are much too diverse—and each one unique—for the homogenized food policy that is likely to result from central government.

The time has come for a "re-org" of our centralized food-production system. Access to healthful food should not be a privilege, but a fundamental right. The current agricultural production methods no longer seem ideal for much of our population. As the demand—and need—for affordable locally produced food rises, it is becoming abundantly clear, from the success stories to date, that the most effective food policies lie not within a central government body, but a local one. If daily access to safe, nutritious, culturally acceptable produce, at little to no cost, is necessary to improve the health and lives of city dwellers, then city government will need to lead the charge. In doing so, it will ultimately have the greatest impact on food security for urban communities. And municipalities will have to be proactive in organizing and implementing a food production system that benefits *all* members in their community. In the very near future, the strength of our country may be determined by the ability of communities to provide food for themselves. As such, underutilized land—under community control—that is arable and available, and that has potential to bring people together in a wholesome manner to grow wholesome food, needs to be explored. Public officials incorporating the use of public space into the programs and policies they craft for a system of public produce, can go a long way toward achieving goals of food security, community health, and economic prosperity— for our cities and our nation.

Acknowledgments

I would first like to express my sincere gratitude to Heather Boyer of Island Press. Not only am I grateful for her insight and editing prowess, but it was Heather's tireless championing of my project to her editorial board that transformed an idea into a published work. Thank you, Heather. Your support is greatly appreciated.

Courtney Lix and Sharis Simonian at Island Press provided quick responses to my many questions, good advice, and fantastic assistance throughout the project. They and the staff at Island Press are courteous, professional, and second to none.

I have had the pleasure of working with Susan Arritt on both of my book projects, and I am extremely grateful to Island Press for assigning this project to her. Susan is a gifted editor with extraordinary talent. And, over the past year, she has become a valued acquaintance. I would do well to have her edit all of my future books.

I am indebted to many municipal officials for their ideas, insight, and inspiration. *Public Produce* would not have come to fruition without my learning from the deeds of these truly dedicated public servants. Many thanks to Teva

Dawson, with the City of Des Moines, and Leslie Pohl-Kosbau, with the City of Portland. At the City of Davenport, Tom Flaherty, Susan Anderson, and Roy De-Witt provided good counsel and inspiration for this book. I am also grateful to Pamela Miner, Matt Flynn, and Dawn Sherman, also from the City of Davenport, for their understanding and support of this extracurricular endeavor.

Stacie Pierce, a member of OPENrestaurant and pastry chef at Chez Panisse, took much-appreciated time from her busy schedule to talk with me about the evolving food and foraging philosophy in the San Francisco Bay Area. She is obviously passionate about cuisine, culture, and the environment, and her commitment to quality, locally sourced food is inspiring. I believe Stacie and her colleagues will soon be setting new standards for how restaurants obtain food and what diners will demand.

This project could not have been completed without the broad aid and comfort of family, especially during the two months preceding the manuscript's deadline. Mary and Dave Nordahl, and Donna and Ken Jensen provided shelter, food, transportation, and welcomed childcare. We disrupted their lifestyles and living arrangements for a few weeks, but I hope the intimate time spent with their grandchildren was just compensation.

Many thanks to my brother, Derek Nordahl, and my "siblings-in-law": Todd Jensen, Greg and Heather McAvoy-Jensen, and Viviana Nordahl. Derek was very helpful—on short notice—with illustrations for the book. Todd provided ample work space, provocative conversation, excellent food, and fine drink (the latter of which stimulated even more provocative conversation). Greg provided needed books for reference, as well as relevant online articles. His wife, Heather, provided considerate support and encouragement, both to me and, more importantly, to my wife. Vivi kept me abreast of all the food headlines and happenings in the Bay Area. She forwarded many articles related to urban agriculture, and was as close to a graduate student researcher that an author outside of academe could hope to have.

Finally, I would like to express my deepest gratitude to my wife, for her unconditional love, support, and enduring friendship. She is truly my biggest champion, and my greatest source of inspiration. Thank you, Lara.

Notes

Introduction

1. Thomas A. Lyson, *Civic Agriculture: Reconnecting Farm, Food, and Community* (Lebanon, NH: University Press of New England, 2004), 21.
2. In 2008, Wendy's launched a new ad campaign that sought to distinguish its menu items from fast food. Though its menu was still heavily comprised of fried hamburgers, French fries, chicken nuggets, soft drinks, and shakes, the new tagline that concluded each commercial was "It's waaaay better than fast food. It's Wendy's." http://www.wendys.com/ads/ (last accessed October 29, 2008).
3. Mark Winne, *Closing the Food Gap: Resetting the Table in the Land of Plenty* (Boston: Beacon Press, 2008), 13–14.
4. Kent Garber, "At Last, Some Respect for Fruits and Veggies," *U.S. News & World Report*, March 13, 2008, Nation section.
5. Holly Hill, "Food Miles and Marketing," National Sustainable Agriculture Information Service, http://attra.ncat.org/attra-pub/foodmiles.html (last accessed July 15, 2008).
6. Lauren Shepherd, "McDonald's 1Q profit rises nearly 4 percent," *The Huffington Post*, April 22, 2009, http://www.huffingtonpost.com/2009/04/22/mcdonalds-profit-rises-ne_n_189923.html (last accessed May 9, 2009).
7. Susan Anderson, in an e-mail message to the author, July 8, 2008.
8. Ibid.
9. Lyson, 63–64.
10. Ibid, 62.
11. Zach Patton, "Fresh Fight," *Governing*, April 2008, 42.

12. The Oxford University Press anointed "locavore" the 2007 Word of the Year, http://blog.oup.com/ 2007/11/locavore/ (last accessed November 1, 2008).

13. Anderson, 2008.

Chapter 1

1. James Howard Kunstler, *The Long Emergency: Surviving the End of Oil, Climate Change, and Other Converging Catastrophes of the Twenty-First Century* (New York: Grove Press, 2005), 239.

2. Michael Pollan, "Farmer in Chief," *New York Times Magazine*, October 12, 2008, New York edition, MM62.

3. Laura J. Lawson, *City Bountiful* (Berkeley: University of California Press, 2005), 171.

4. Ibid., 115.

5. Jennifer Steinhaur, "Governor Declares Drought in California," *New York Times*, June 5, 2008, U.S. section.

6. Amy Lorentzen, "River Slowly Dropping, but Iowa Still Flooded," Associated Press, http://news .yahoo.com/s/ap/20080614/ap_on_re_us/midwest_flooding (last accessed July 4, 2008).

7. Juanita Kakalec, in a conversation with the author, March 23, 2008.

8. Daniel Solomon, *Global City Blues* (Washington, DC: Island Press, 2003), 17.

9. Winne, 14.

10. Rich Pirog et al., *Food, Fuel, and Freeways: An Iowa perspective on how far food travels, fuel usage, and greenhouse gas emissions* (Ames, IA: Leopold Center for Sustainable Agriculture, Iowa State University, 2001). An interesting graphic associated with the report, which illustrates how far selected produce varieties travel by truck to reach a terminal market in Chicago, can be viewed at http://www .leopold.iastate.edu/pubs/staff/ppp/produce_chart.html (last accessed May 10, 2009).

11. Sue Weaver, "Origins of the Orchard," *Orcharding* 8(2008): 8.

12. Pirog, 6.

13. Data on Iowa's commodity crops and cash receipts were obtained from the U.S. Department of Agriculture's Economic Research Service. Data sets for all commodities by state can be obtained from http://www.ers.usda.gov/data/farmincome/FinfidmuXls.htm (last accessed February 7, 2009).

14. For more information about the FDA's Produce Safety Action Plan, visit http://www.cfsan.fda.gov/ ~dms/prodpla2.html (last accessed February 3, 2009).

15. Center for Food Safety and Applied Nutrition, "Lettuce Safety Initiative," *U.S. Food and Drug Administration* (August 23, 2006): http://www.cfsan.fda.gov/%7Edms/lettsafe.html (last accessed February 3, 2009).

16. For a documented history of the *E. coli* outbreak in spinach, including state-by-state statistics, visit the CDC web site: http://www.cdc.gov/ecoli/2006/september/updates/ (last accessed May 10, 2009).

17. U.S. Food and Drug Administration/Center for Food Safety and Applied Nutrition, "Leafy Greens Safety Initiative—2nd year," October 4, 2007, http://www.cfsan.fda.gov/~dms/lettsaf2.html (last accessed February 3, 2009).

18. Information on the outbreak of *Salmonella* Saintpaul-tainted peppers, including state-by-state statistics for this particular outbreak, can be found on the CDC web site: http://www.cdc.gov/ salmonella/saintpaul/jalapeno/ (last accessed May 10, 2009).

19. Information on the outbreak of *Salmonella* Typhimurium, including state-by-state statistics for this

serotype, can be found on the CDC website: http://www.cdc.gov/salmonella/typhimurium/ update.html (last accessed May 10, 2009).

20. Ricardo Alonso-Zaldivar, "Private inspections of food companies seen as weak," Associated Press, March 20, 2009, http://news.yahoo.com/s/ap/20090320/ap_on_go_co/salmonella_outbreak (last accessed March 20, 2009).

21. Ricardo Alonso-Zaldivar and Brett J. Blackledge, "Stewart Parnell, Peanut Corp Owner, Refuses To Testify To Congress In Salmonella Hearing," Associated Press, February 11, 2009, http://www .huffingtonpost.com/2009/02/11/stewart-parnell-peanut-co_n_166058.html (last accessed May 10, 2009).

22. Information on the outbreak of *Salmonella* Saintpaul-tainted alfalfa, including state-by-state statistics for this particular outbreak, can be found on the CDC website: http://www.cdc.gov/salmonella/ saintpaul/alfalfa/ (last accessed May 10, 2009).

23. Information on the outbreak of *Salmonella*-tainted pistachios can be found on the CDC website: http://www.cdc.gov/salmonella/pistachios/update.html (last accessed May 10, 2009).

24. Results from the AP-Ipsos poll were gleaned from Ricardo Alonso-Zaldivar, "Food Safety Worries Change Buying Habits," ABC News, July 18, 2008, http://a.abcnews.com/US/WireStory?id= 5401116&page=1 (last accessed January 29, 2009).

25. Ibid.

26. Marion Nestle, in an e-mail exchange with the author, October 15, 2008.

27. Pollan, "Farmer in Chief."

28. Amanda Gardner, "FDA Expands Tomato Warning Nationwide," *HealthDayNews*, June 10, 2008, http://www.healthday.com/Article.asp?AID=616391 (last accessed November 28, 2008).

29. Centers for Disease Control and Prevention, "Outbreak of *Salmonella* Serotype Saintpaul Infections Associated with Multiple Raw Produce Items—United States, 2008," *Morbidity and Mortality Weekly Report* 57, no. 34 (August 29, 2008), 929–934.

30. Centers for Disease Control and Prevention/Division of Foodborne, Bacterial and Mycotic Diseases, "Salmonellosis," http://www.cdc.gov/nczved/dfbmd/disease_listing/salmonellosis_gi.html#6 (last accessed January 27, 2009).

31. World Health Organization, *Guidelines for the Safe Use of Wastewater, Excreta and Greywater*, vol. IV of Excreta and Greywater Use in Agriculture (Geneva: WHO Press, 2006), http://whqlibdoc.who .int/publications/2006/9241546859_eng.pdf.

32. Marion Nestle, *Safe Food: Bacteria, Biotechnology, and Bioterrorism* (Berkeley: University of California Press, 2003), 59.

33. Ibid.

34. As reported in Pollan, "Farmer in Chief."

35. Ibid.

36. Kristin Collins, "Grower settles with limbless child," *News & Observer* (Raleigh, NC), March 25, 2008, B1.

37. Information regarding lead contamination in soil and remediation strategies, including specific quotes, was gleaned from Carl J. Rosen, "Lead in the Home Garden and Urban Soil Environment," University of Minnesota Extension, 2002, http://www.extension.umn.edu/distribution/ horticulture/DG2543.html (last accessed January 29, 2009).

38. Colin Beavan, "The No Impact Experiment," The No Impact Man Blog, entry posted February 21,

2007, http://noimpactman.typepad.com/blog/2007/02/the_no_impact_e.html (last accessed February 7, 2009).

39. Colin Beaven, "Like falling off a log," The No Impact Man Blog, entry posted March 21, 2008, http://noimpactman.typepad.com/blog/2008/03/index.html (last accessed February 7, 2009).

40. Michael Pollan, *The Omnivore's Dilemma: A Natural History of Four Meals* (New York: The Penguin Press, 2006), 100–108.

41. Michael Pollan, *In Defense of Food: An Eater's Manifesto* (New York: The Penguin Press, 2008), 158.

42. Nutrition information for peaches obtained from: Frances Sizer and Eleanor Whitney, *Nutrition: Concepts and Controversies*, 6th ed. (Minneapolis/St. Paul: West Publishing Company, 1994), A-20. Nutrition information for McDonald's double cheeseburger and other menu items available at: http://nutrition.mcdonalds.com (last accessed February 2, 2009).

43. Winne, xvi–xvii.

44. The reporting on Huntington's health problems and the quotes have been gleaned from Mike Stobbe, "West Virginia town called 'unhealthiest city' in nation," *Quad-City Times* (Davenport, IA), November 17, 2008, A1.

45. Statistics from the Los Angeles County Department of Public Health, as reported in Associated Press, "L.A. OKs moratorium on fast-food restaurants," MSNBC, July 29, 2008, http://www.msnbc.msn.com/id/25896233/ (last accessed February 2, 2009).

46. Ibid.

47. Kim Severson, "Los Angeles Stages a Fast Food Intervention," *New York Times*, August 13, 2008, F1.

48. Data obtained from the Centers for Disease Control and Prevention's Behavioral Risk Factor Surveillance System. Obesity is defined as having a Body Mass Index of 30 or higher.

49. Centers for Disease Control and Prevention, *Overweight and Obesity*, http://www.cdc.gov/nccdphp/dnpa/obesity/childhood/index.htm, (last accessed May 10, 2009).

50. Centers for Disease Control and Prevention, *Chronic Disease Prevention and Health Promotion*, http://www.cdc.gov/nccdphp/publications/aag/dnpa.htm, (last accessed May 10, 2009).

51. Centers for Disease Control and Prevention, "Preventing Obesity and Chronic Diseases Through Good Nutrition and Physical Activity," http://www.cdc.gov/nccdphp/publications/factsheets/Prevention/obesity.htm (last accessed February 7, 2009).

52. Maring's farmers' markets have proven so popular that Kaiser Permanente created a web page for people to find their nearest hospital farmers' market: http://members.kaiserpermanente.org/redirects/farmersmarkets. Also of note, Kaiser Permanente hosts another web page devoted to Maring's recipes for healthy food. His "Barackoli with garlic and lemon," a tribute to President Barack Obama, is particularly intriguing: http://recipe.kaiser-permanente.org/kp/maring/2009/01/barackoli_with_garlic_and_lemo.php. The reporting of the history of Kaiser's farmers' markets and the quote by Maring are from Project for Public Spaces, "Kaiser Farmers Markets," http://www.pps.org/markets/info/market_profiles/food_insecurity/kaiser (last accessed January 29, 2009).

53. Michelle D. Florence et al., "Diet Quality and Academic Performance," *Journal of School Health* 78 (2008): 212.

54. Severson.

55. Eric Schlosser, *Fast Food Nation* (New York: Perennial, 2002), 42–46.

56. Pollan, "Farmer in Chief."

Chapter 2

1. Jens Jensen, "A Greater West Side Park System," *West Chicago Park Commissioners* (1920).

2. William H. Whyte, *The Social Life of Small Urban Spaces* (New York: Project for Public Spaces, 1980), 50–53.

3. From a report prepared for the Providence Urban Agriculture Policy Task Force by Benjamin Morton, "Planning for Appropriately Scaled Agriculture in Providence," Fall 2006, http://www.farm freshri.org/learn/docs/urbanag-planning.pdf (last accessed December 28, 2008).

4. Seattle's use of the term "urban villages" generally refers to a densely populated collection of mixed-use neighborhoods that also may be regional centers. For more information on urban agriculture and urban villages, see The City of Seattle, *Comprehensive Plan: Toward a Sustainable Seattle* (Seattle: City of Seattle, January 2005). The quoted community gardening goal was obtained from one of the Plan's Appendices: "Urban Village Element."

5. Michael Grunwald, "Mayor Giuliani Holds a Garden Sale," *Washington Post*, May 12, 1999, A1.

6. This quote, and the information about New York's garden auction, was culled from Laura J. Lawson, *City Bountiful* (Berkeley: University of California Press, 2005), 260–263.

7. Jerry Kaufman and Martin Bailkey, "Farming Inside Cities: Entrepreneurial Urban Agriculture in the United States," working paper, Lincoln Institute of Land Policy, 2000, 85.

8. Bacon's quote was originally recorded in Rose DeWolf, "We Look to the Future . . . And Learn from our Past," *Philadelphia Daily News*, July 24, 2000, 05.

9. For more information on the National Vacant Properties Campaign, visit: http://www.vacant properties.org (last accessed February 8, 2009).

10. John Gallagher, "Acres of barren blocks offer chance to reinvent Detroit," *Detroit Free Press*, December 15, 2008, http://www.freep.com/article/20081215/NEWS01/812150342 (last accessed February 8, 2009).

11. For more information on land trusts and conservation opportunities, visit Land Trust Alliance, www.lta.org or The Trust for Public Land, www.tpl.org.

12. NeighborSpace, "How We Do It," http://neighbor-space.org/howwedoit.htm (last accessed January 15, 2009).

13. Kaufman and Bailkey, 30–31.

14. Luc J. A. Mougeot, *Growing better Cities: Urban agriculture for sustainable development* (Ottawa, ON: International Development Research Centre, 2006), 64.

15. Ibid, 64–65.

16. Plans for the City of Des Moines' myriad edible landscapes and the goals behind those plans can be viewed at: http://www.ci.des-moines.ia.us/departments/pr/Comm_Gard/digging_deeper.htm# landscapes (last accessed January 21, 2009).

17. Kathleen E. Dickhut, et al., "Chicago: Eat Local Live Healthy," *City of Chicago Department of Planning and Development* (2006), 2.

18. Information about the City of Chicago's honey was obtained from Veronica Hinke, "The Bee Line: The story behind City Hall's 'rooftop honey,'" *Newcity Chicago*, November 07, 2006, Food & Drink section; an e-mail from Michael Thompson, beekeeper, to the author, December 30, 2008; and a conversation with City of Chicago Cultural Center staff, January 13, 2009.

19. Michael Schacker, *A Spring without Bees: How Colony Collapse Disorder Has Endangered Our Food Supply* (Guilford, CT: The Lyons Press, 2008), 2.

20. Dickhut et al., 1
21. Ibid.

Chapter 3

1. Pollan, *In Defense of Food*, 184.
2. Associated Press, "Farm's Open Harvest Draws 40,000 in Colorado," *New York Times*, November 24, 2008, A14. Also from a broadcast interview of Joe Miller, "In Colo., Veggie Giveaway Spurs Massive Response," *All Things Considered*, National Public Radio, November 24, 2008, http://www.npr.org/templates/story/story.php?storyId=97418921 (last accessed December 13, 2008).
3. Agnès Varda, *The Gleaners and I* (France: Cinè-Tamaris, 2000), DVD.
4. Pollan, "Farmer in Chief."
5. Robert Steuteville, "Old McDonald had an organic TND?" *New Urban News*, December 2008, 13.
6. Ibid.
7. Pollan, *The Omnivore's Dilemma*, 397.
8. David Burns, Matias Viegener, and Austin Young, "What is Fallen Fruit?" http://www.fallenfruit.org/whatisfallenfruit.html (last accessed August 2, 2008).
9. Winne, 32–33.
10. Gleaners, "Hunger in Indiana," http://www.gleaners.org/hunger_indiana.shtm (last accessed December 14, 2008).
11. Statistics for hunger in San Francisco were obtained from the San Francisco Food Bank website (http://www.sffoodbank.org), and from *A Look at Hunger in San Francisco: Neighborhood Profiles of Hunger and Food Programs* (San Francisco: San Francisco Food Bank, July 2007), http://www.sffoodbank.org/about_hunger/local_study.pdf.
12. Marguerite Nowak, food advocacy manager of the San Francisco Food Bank, in a telephone conversation with the author, February 9, 2009.
13. Stacie Pierce, in a telephone conversation with the author, January 28, 2009.
14. Winne, 55.

Chapter 4

1. Richard Register, *Ecocity Berkeley: Building Cities for a Healthy Future* (Berkeley: North Atlantic Books, 1987), 41–42.
2. Though commonly considered a tropical plant, the "banana" variety of passion fruit (*Passiflora mollissima*) is more vigorous and tolerant of frost than the typical commercial varieties, making this cultivar suitable to many areas in the United States.
3. Register, 42–43.
4. Understandably, the olive oil program at UC Davis has received considerable media attention. The information gathered for this book was gleaned from the many news articles posted on the university's website: http://oliveoil.ucdavis.edu/about_news.shtml. In particular, see Jim Downing, "Tasting Success: UCD celebrates another year of turning its olive mess into a moneymaker," *The Sacramento Bee*, March 20, 2008, Business section.
5. A Business Improvement District (BID) is a type of public/private partnership, whereby private

businesses in a particular area (such as downtown, or along a neighborhood retail street) elect to pay an additional tax, which is used to fund improvements serving all the businesses. These improvements typically include improvements to the public realm (e.g., sidewalks, decorative street lighting, banners, trees, landscaping, and other streetscape enhancements).

6. "P-Patch" is a term unique to Seattle, and is used as a colloquialism for community garden plots. The "P" is short for Picardo, the family that, in the early 1970s, lent their farm to the City of Seattle for community gardening. For more information on the fascinating history of Seattle's P-Patch program, visit the City's website: http://www.seattle.gov/Neighborhoods/ppatch/history.htm (last accessed May 17, 2009).

7. Kaufman and Bailkey, 78.

8. If the lawns are regularly treated with chemical fertilizers and pesticides, consider if irrigation run off would pose a risk to edibles. Also, it should be noted that reclaimed water is increasingly being used for irrigation of ornamental landscapes by many municipalities, particularly in California. While there are ongoing debates over the use of reclaimed water for irrigating food crops, it is generally advised to avoid spraying edible parts of plants directly with reclaimed water.

9. Bill Mollison, *Permaculture: A Designers' Manual* (Tyalgum, Australia: Tagari Publications, 1988), ix.

10. Teva Dawson, in a telephone conversation with the author, November 14, 2008.

11. The quotes and information for the Jamaica, Queens, community garden was obtained from Anne Raver, "Healthy Spaces, for People and the Earth," *New York Times*, November 6, 2008, D6.

12. Portland Fruit Tree Project, "About the Project," http://portlandfruit.org/WebPages/About.html (last accessed December 28, 2008).

Chapter 5

1. Victory Seed Company, "A History of Victory Gardening," http://www.victoryseeds.com/TheVictory Garden/page5.html (last accessed December 14, 2008).

2. Pollan, *In Defense of Food*, 149.

3. Eric Schlosser, *Fast Food Nation* (New York: Perennial, 2002), 262.

4. Lawson, 202.

5. http://cuesa.org/.

6. OPENrestaurant, "A Project," http://openrestaurant.org/project-description (last accessed February 13, 2009).

7. Stacie Pierce, pasty chef at Chez Panisse and organizer of OPENrestaurant, in a telephone conversation with the author, January 28, 2009.

8. OPENrestaurant, "OPENrestaurant at the Yerba Buena Center for the Arts," http://openrestaurant .org/2008/12/31/openrestaurant-at-the-yerba-buena-center-for-the-arts (last accessed February 13, 2009).

9. Mustapha Koc et al., eds., *For Hunger-Proof Cities: Sustainable Urban Food Systems* (Ottawa, ON: International Development Research Centre, 1999), 58.

10. Winne, 140.

11. Sheldon Margen and the Editors of the University of California at Berkeley *Wellness Letter, The Wellness Encyclopedia of Food and Nutrition: How to Buy, Store, and Prepare Every Variety of Fresh Food* (New York: Rebus, 1992), 40.

12. Teva Dawson, in an e-mail message to the author, December 18, 2008.
13. Lawson, 51.
14. Pollan, "Farmer in Chief."
15. The Edible Schoolyard, "Mission," http://www.edibleschoolyard.org/mission-goals (last accessed May 19, 2009).
16. Harvard University Dining Services, "The Food Literacy Project," http://www.dining.harvard.edu/flp/index.html (last accessed February 28, 2009).
17. Marguerite Nowak, in a telephone conversation with the author, February 9, 2009.
18. Text is from the maps accessible at: http://www.fallenfruit.org/maps.html (last accessed February 14, 2009).

Conclusion

1. Cleo Woelfle-Erskine, ed., *Urban Wilds: Gardeners' Stories of the Struggle for Land and Justice* (Oakland: Water/Under/Ground, 2003), 3.
2. Lawson, 171.
3. www.sfvictorygardens.org (last accessed February 22, 2009).
4. Quoted from Lawson, 175. These goals and other proceedings from the National Defense Gardening Conference are summarized in U.S. Office of Civilian Defense, *Garden for Victory: Guide for Planning the Local Victory Garden Program* (Washington, DC: U.S. Government Printing Office, 1943).
5. Winne, 149–150.
6. Schlosser, 266.
7. Lawson, xiv.
8. Mougeot, vi.
9. Lyson, 76–77.
10. Winne, 54–55.
11. Lisa Van Cleef, "Gardening Conquers All: How to cut your jail recidivism rates by half," *SF Gate*, December 18, 2002, http://www.sfgate.com/cgi-bin/article.cgi?file=%2Fgate%2Farchive%2F2002%2F12%2F18%2Fgreeng.DTL (last accessed February 23, 2009).
12. Jane Gross, "A Jail Garden's Harvest: Hope and Redemption," *New York Times*, September 3, 1992, U.S. section.
13. Winne, 62.
14. Green for All, "The Vision," http://www.greenforall.org/about-us/the-vision (last accessed March 4, 2009).
15. Dickhut, 10.
16. Des Moines Parks Department, "The Des Moines Community Garden Coalition," City of Des Moines, http://www.ci.des-moines.ia.us/departments/pr/Comm_Gard/ (last accessed December 21, 2008).
17. Pollan, "Farmer in Chief.
18. Marian Burros, "Obamas to Plant Vegetable Garden at White House," *New York Times*, March 20, 2009, A1.
19. Office of the Press Secretary, "WEEKLY ADDRESS: President Barack Obama Announces Key FDA Appointments and Tougher Food Safety Measures," *The White House*, March 14, 2009. Minutes

of the presidential address, along with the full remarks, can be viewed at http://www.white house.gov/the_press_office/Weekly-Address-President-Barack-Obama-Announces-Key-FDA-Appointments-and-Tougher-F/ (last accessed May 22, 2009).

20. Hayley Boriss and Heinrich Brunke, "Commodity Profile: Broccoli," *Agricultural Issues Center, University of California* (December 2005): 1–2.

Bibliography

Books

Brown, Lester R. *Plan B: Rescuing a Planet under Stress and a Civilization in Trouble*. New York: W. W. Norton, 2003.

Edinger, Philip. *Herbs: An Illustrated Guide*. Menlo Park, CA: Sunset Publishing, 1996.

Herr, Serena. *Trees for San Francisco: A Guide to Street-Tree Planting and Care*. San Francisco: Friends of the Urban Forest, 1995.

Koc, Mustapha, Rod McRae, Jennifer Walsh, and Luc J. A. Mougeot, eds. *For Hunger-Proof Cities: Sustainable Urban Food Systems*. Ottawa, ON: International Development Research Centre, 1999.

Kunstler, James Howard. *The Long Emergency: Surviving the End of Oil, Climate Change, and Other Converging Catastrophes of the Twenty-First Century*. New York: Grove Press, 2005.

Lawson, Laura J. *City Bountiful: A Century of Community Gardening in America*. Berkeley: University of California Press, 2005.

Lyson, Thomas A. *Civic Agriculture: Reconnecting Farm, Food, and Community*. Medford, MA: Tufts University Press, 2004.

Margen, Sheldon, and the editors of the University of California *Wellness Letter*. *The Wellness Encyclopedia of Food and Nutrition: How to Buy, Store, and Prepare Every Variety of Fresh Food*. New York: Rebus, 1992.

Mollison, Bill. *Permaculture: A Designers' Manual*. Tyalgum, Australia: Tagari Publications, 1988.

———. *Introduction to Permaculture*. Tyalgum, Australia: Tagari Publications, 1991.

Mougeot, Luc J.A. *Growing Better Cities: Urban Agriculture for Sustainable Development.* Ottawa, ON: International Development Research Centre, 2006.

Nestle, Marion. *Safe Food: Bacteria, Biotechnology, and Bioterrorism.* Berkeley: University of California Press, 2003.

Pollan, Michael. *The Omnivore's Dilemma: A Natural History of Four Meals.* New York: The Penguin Press, 2006.

———. *In Defense of Food: An Eater's Manifesto.* New York: The Penguin Press, 2008.

Register, Richard. *Ecocity Berkeley: Building Cities for a Healthy Future.* Berkeley: North Atlantic Books, 1987.

———. *Ecocities: Rebuilding Cities in Balance with Nature.* Gabriola Island, BC: New Society Publishers, 2006.

Schacker, Michael. *A Spring without Bees: How Colony Collapse Disorder Has Endangered Our Food Supply.* Guilford, CT: The Lyons Press, 2008.

Schlosser, Eric. *Fast Food Nation: The Dark Side of the All-American Meal.* New York: Perennial, 2002.

Sizer, Frances and Eleanor Whitney. *Nutrition: Concepts and Controversies*, 6th ed. Minneapolis/St. Paul: West Publishing, 1994.

Solomon, Daniel. *Global City Blues.* Washington, DC: Island Press, 2003.

Walheim, Lance. *Vegetable Gardening.* Menlo Park, CA: Sunset Publishing, 1998.

Waters, Alice. *Chez Panisse Café Cookbook.* New York: HarperCollins, 1999.

Whyte, William H. *The Social Life of Small Urban Spaces.* New York: Project for Public Spaces, 1980.

———. *City: Rediscovering the Center.* New York: Doubleday, 1988.

Winne, Mark. *Closing the Food Gap: Resetting the Table in the Land of Plenty.* Boston: Beacon Press, 2008.

Woelfle-Erskine, Cleo, ed. *Urban Wilds: Gardeners' Stories of the Struggle for Land and Justice.* Oakland: Water/Under/Ground, 2003.

Film

Varda, Agnès. *The Gleaners and I.* France: Cinè-Tamaris, 2000. DVD.

Papers, Reports, and Articles

Florence, Michelle D., Mark Asbridge, and Paul J. Veugelers. "Diet Quality and Academic Performance." *Journal of School Health* 78 (2008), 209–215.

Kaufman, Jerry and Martin Bailkey. "Farming Inside Cities: Entrepreneurial Urban Agriculture in the United States." Working paper, Lincoln Institute of Land Policy, 2000.

Pirog, Rich, Timothy Van Pelt, Kamyar Enshayan, and Ellen Cook. *Food, Fuel, and Freeways: An Iowa perspective on how far food travels, fuel usage, and greenhouse gas emissions.* Ames, IA: Leopold Center for Sustainable Agriculture, Iowa State University, 2001.

Pollan, Michael. "Farmer in Chief." *New York Times Magazine*, October 12, 2008.

Selected Internet resources promoting food education and literacy

100-Mile Diet: Local Eating for Global Change (http://100milediet.org/)

American Community Gardening Association (http://communitygarden.org)

Center for Urban Education about Sustainable Agriculture (http://cuesa.org/)

Centers for Disease Control and Prevention (http://cdc.gov/)

Eat Local Challenge (http://eatlocalchallenge.com/)

Feeding America (formerly America's Second Harvest) (http://feedingamerica.org/)

Fallen Fruit (http://fallenfruit.org/)

Food Politics (http://www.foodpolitics.com/)

Forage Oakland (http://forageoakland.blogspot.com/)

Harvard University Dining Services: The Food Literacy Project (http://dining.harvard.edu/flp/)

Key Ingredients: America by Food (http://keyingredients.org/)

No Impact Man (http://noimpactman.typepad.com/)

OPENrestaurant (http://openrestaurant.org/)

Slow Food International (http://slowfood.com/)

The Edible Schoolyard (http://www.edibleschoolyard.org/)

U.S. Department of Agriculture Economic Research Service (http://ers.usda.gov/)

Victory Seed Company (http://victoryseeds.com/)

Index